Speed

Vaclav Smil is Distinguished Professor Emeritus at the University of Manitoba. He is the author of more than forty books on topics including energy, environmental and population change, food production and nutrition, technical innovation, risk assessment and public policy. His books for Viking/Penguin, including the *New York Times* bestseller *How the World Really Works*, have been published in over thirty languages. No other living scientist has had more books (on a wide variety of topics) reviewed in the leading scientific journal *Nature*. A Fellow of the Royal Society of Canada, in 2010 he was named by *Foreign Policy* as one of the Top 100 Global Thinkers.

By the same author

Speed

How It Explains the World

VACLAV SMIL

PENGUIN
VIKING

VIKING

UK | USA | Canada | Ireland | Australia
India | New Zealand | South Africa

Viking is part of the Penguin Random House group of companies
whose addresses can be found at global.penguinrandomhouse.com

Penguin Random House UK,
One Embassy Gardens, 8 Viaduct Gardens, London s w 11 7 b w

penguin.co.uk

Penguin
Random House
UK

First published 2025
001

Set in 12/14.75pt Bembo Book MT Pro
Typeset by Six Red Marbles UK, Thetford, Norfolk
Printed and bound in Great Britain by Clays Ltd, Elcograf S.p.A.

The authorized representative in the EEA is Penguin Random House Ireland,
Morrison Chambers, 32 Nassau Street, Dublin D02 Y H68

A CIP catalogue record for this book is available from the British Library

ISBN: 978-0-241-75454-2

Penguin Random House is committed to a sustainable future
for our business, our readers and our planet. This book is made from
Forest Stewardship Council® certified paper.

MIX
Paper | Supporting
responsible forestry
FSC® C018179

Contents

Preface: From Speed's Absence to Singularity

Writing interdisciplinary books is an inherently risky endeavor. Given the extent of our understanding and the speed at which it is being expanded, no amount of diligence will ensure that an author can be equally confident in dealing with all subjects that fall within the remits of wide-ranging topics. There are many ways to minimize errors, omissions, and questionable interpretations while retaining the cohesion that an author tries to impart to a single-volume treatment of a mighty topic, and one of them I pursue diligently even before I begin to write: I always prefer to get a reasonably close overview of what has gone before, by consulting books that have been published on the topic in order to ascertain what has been well covered, what has been only cursorily treated, and what might be missing.

Speed is, obviously, one of the most important concerns of modern civilization: we have devoted enormous resources in order to produce everything faster (think of large combines harvesting wheat, or robotic car assembly lines welding chassis), to travel everywhere in shorter times (intercontinental flights that, with a little help from tailwinds, travel at the speed of sound), and to communicate instantly with everybody anywhere (smart mobile phones have made that a mundane affair). Hence it is quite surprising that speed has been a relatively poorly covered variable. For a quick proof of recent scientific interest, compare the numbers of articles listed in PubMed (a freely accessible archive of biomedical and life-science literature) at the end of 2024 dealing with growth (another modern obsession) and speed: the respective totals were about 3 million and about 250,000 items.[1] Web of Science, covering all scientific disciplines (and accessible only via institutional subscriptions), gives, of course, a far more comprehensive account, but one that is similarly (albeit less extremely) skewed: at the end of 2024 it listed 4.6 million entries dealing (directly and tangentially) with growth and 1.3 million with speed.[2]

I thought that the coverage of speed in famous classic texts, and even more so in modern interdisciplinary books, would be different—but I was struck by the curious absence of not only an explicit but even a peripheral interest. The list of antique authors who had nothing to say about speed includes the compilers of the New Testament, Cicero (*Orations*), and Marcus Aurelius (*Meditations*); other kindred ancient writings do not contain a single mention of speed (in whatever context). This *magnum silentium* extends through the Middle Ages (Thomas Aquinas, *Summa Theologica*; the *Rubaiyat* of Omar Khayyam) to the early modern era (John Locke's *An Essay Concerning Human Understanding*; Robert Malthus's *An Essay on the Principle of Population*) and into the 19th century: nothing in Thoreau (*Walden*) or Emerson (*Nature*), and Walt Whitman's few mentions of speed are (not surprisingly) concentrated in the self-centered "Song of Myself."

I would not argue that checking book indexes is an unfailingly revealing way to gauge the extent of interest in a particular topic, but discovering a widespread absence of specific attention to what is undoubtedly one of the dominant properties of physical processes—the evolution of life—and the development of civilizations is a noteworthy reality. This could become a tedious recital of well-known titles, and that is why I will limit the list just to those volumes where I found the lack of an explicit reference to anything to do with speed most surprising.

In life sciences, they include Charles Darwin's *On the Origin of Species* (1859), D'Arcy Wentworth Thompson's classic *On Growth and Form* (the first edition in 1917), and Geoffrey West's *Scale* (2017).[3] In history and social sciences, Jürgen Osterhammel's nearly 1,000-page-long overview of the 19th century (*The Transformation of the World*, in German in 2009, in English in 2014), George Kingsley Zipf's *Human Behavior and the Principle of Least Effort* (1949), and Joel Mokyr's *Culture of Growth* (2017).[4] In books about engineering, speed gets no special attention either—not in *The Oxford Handbook of Engineering and Technology in the Classical World* (edited by John Peter Oleson in 2008) or in *Makers of the Microchip* by Christoph Lécuyer and David C. Brock

(2010)—and in energy studies it is ignored in Howard Odum's *Environment, Power, and Society* (1971) and in Robert Ayres's *Energy, Complexity and Wealth Maximization* (2016).[5]

Perhaps even more remarkably, explicit attention to speed in its many manifestations is absent even in encyclopedias. On my library shelves I have six sets to which I contributed entries—*Encyclopedia of Energy, Berkshire Encyclopedia of World History, Encyclopedia of the Age of the Industrial Revolution* (the age defined by increasing speed!), *Encyclopedia of Population, Encyclopedia of World Environmental History*, and *Encyclopedia of Nonlinear Science*—and none of their indexes contains a single reference to speed.

The most notable (and expected) exception is Richard Feynman's de facto encyclopedia, his famous three-volume set of *Lectures on Physics* (originally published in 1963) that amounts to a thorough overview of the discipline and whose section on speed opens with a characteristic Feynmanian caveat:

> Even though we know roughly what "speed" means, there are still some rather deep subtleties; consider that the learned Greeks were never able to adequately describe problems involving velocity. The subtlety comes when we try to comprehend exactly what is meant by "speed."[6]

Speed—in fundamental physics this means length (distance) divided by time (L/T)—is a scalar, a physical quantity that is completely described by its magnitude (as is size or time). Velocity, however (*velox* means "swift" or "rapid" in Latin), is a vector because it has both magnitude *and* direction: one might say that velocity is direction-aware speed. When people make comments about a fast-moving car or swiftly incoming tide or rapidly falling meteorite, they do not, strictly speaking, refer to speed, but to near-instantaneous velocities of the observed phenomena—to short-term movements of objects in specific directions. Understanding these fundamentals is requisite for appreciating the modern preoccupation with speed in general and with acceleration in particular.

Acceleration is a vector defined in physics as the rate of change of velocity of an object with respect to time, that is $(L/T)/T = LT^{-2}$, but in historical writings the term has been used not for objects but, overwhelmingly, to describe faster rates of technical, economic, and social change. These accelerations became obvious during the latter half of the 19th century, and many of them have appreciably intensified since the end of the Second World War. After quantifying numerous technical advances of the late 19th century, Henry Adams (great-grandson of John Adams, America's second president) was among the first observers to write about "a new centre, or preponderating mass, artificially introduced on earth in the midst of a system of attractive forces that previously made their own equilibrium, and constantly induced to accelerate its motion till it shall establish a new equilibrium."[7]

In 1919 he went further, dividing past advances into the Mechanical Phase (1600–1900, dominated by animal and human labor and, after 1850, by steam power) and the Electric Phase (electric lights and motors ascendant, with a span of just 17.5 years), which would pass into the Ethereal Phase (lasting just over 4 years) "and bring Thought to the limit of its possibilities in the year 1921. It may well be! Nothing whatever is beyond the range of possibility; but even if the life of the previous phase, 1600–1900, were extended another hundred years, the difference to the last term of the series would be negligible. In that case, the Ethereal Phase would last till about 2025."[8]

But in 2025 we were not, obviously, in the era where "nothing whatever is beyond the range of possibility." The three decades after 1919 brought an unprecedented combination of advances and retreats, as post–First World War hardships, the great economic crisis of the 1930s, and the Second World War demanded more immediate attention on matters of survival, even as major technical innovations continued.

The first post–First World War decade brought accelerated mass adoptions of industrial motors, cars, and radios; new plastics, the first practical gas turbines, and the discovery of nuclear fission came during the 1930s; and the Second World War led to antibiotics, jet engines, nuclear bombs, and the first electronic computers.[9] The rising performance of these innovations led to expectations of further fabulous computational advances, and John von Neumann,

Hungarian-American mathematician and one of the creators of the modern electronic era, might have been the first scientist to posit the arrival of singularity—the end of history as we have known it. His pupil and friend Stanisław Ulam, a Polish-American mathematician and physicist, recalled in von Neumann's obituary how (sometime in the late 1940s) "one conversation centered on the ever-accelerating progress of technology and changes in the mode of human life, which gives the appearance of approaching some essential singularity in the history of the race beyond which human affairs, as we know them, could not continue."[10]

Around the same time, in France, François Meyer and Daniel Halévy wrote about *l'accélération évolutive* and *l'accélération de l'histoire* (evolutionary and historical acceleration), and during the second half of the 20th century came a wave of American writings on accelerated development (with the emphasis on the advances in computing) that included essays and books by Richard Feynman, Gordon Moore, Gerard Piel, Hans Moravec, Vernor Vinge, and Richard Coren.[11]

Building on these intellectual foundations, in 2001 Ray Kurzweil, an American computer scientist, formulated his law of accelerating returns:

> An analysis of the history of technology shows that technological change is exponential, contrary to the common-sense "intuitive linear" view. So we won't experience 100 years of progress in the 21st century—it will be more like 20,000 years of progress (at today's rate) . . . There's even exponential growth in the rate of exponential growth. Within a few decades, machine intelligence will surpass human intelligence, leading to The Singularity—technological change so rapid and profound it represents a rupture in the fabric of human history. The implications include the merger of biological and nonbiological intelligence, immortal software-based humans, and ultra-high levels of intelligence that expand outward in the universe at the speed of light.[12]

In 2005, in *The Singularity Is Near*, Kurzweil offered a firm date for the event: it is to come in 2045.[13]

Subsequent publications included the examinations of some

sociological aspects of modern manifestations of speed, but perhaps the most important outcome of examining the Great Acceleration—the process that has affected everything from global population growth to mass-scale urbanization, from the synthesis of chemical fertilizers to emissions of CO_2, and from fossil fuel combustion to the expansion of shrimp aquaculture—has been the conclusion that these transformations have already altered the Earth's biosphere to such an extent that we are now living in the Anthropocene, a new geological era dominated by human actions.[14]

One modern book with speed (rather than acceleration) in its title is a brief volume, *Vitesse et Politique* (*Speed and Politics*), published in 2006 by Paul Virilio, a French philosopher and urbanist—but at its very beginning he makes it clear that he is not concerned with speed in standard physical terms but as "Time saved in the most absolute sense of the word, since it becomes human Time directly torn from Death . . ."[15] Then he assures us that "speed as a pure idea without content comes from the sea like Venus, and when Marinetti cries that the universe has been enriched by a new beauty, the beauty of speed, and opposes the racecar to the Winged Victory of Samothrace, he forgets that he is really talking about the same esthetic: the esthetic of the transport engine."[16]

And right on the next page he concludes (working with the two terms he created from *dromos*, the Greek word for racecourse, as an epitome of speed) that "there was no 'industrial revolution,' but only a 'dromocratic revolution'; there is no democracy, only dromocracy; there is no strategy, only dromology . . . It is speed as the nature of dromological progress that ruins progress; it is the permanence of the war of Time that creates total peace, the peace of exhaustion."[17] I readily confess that in this book I do not aspire to such exalted levels of mythologization, generalization, transference, and dromological speculation.

These, then, are the facts. There is no crowded field of comprehensive, synthesizing, realistic writings about speed as a notable variable in natural and human history—even as diverse speeds have always been critical determinants of planetary and organismic evolution, as well as part of our thinking (from ancient legends to modern fiction) and of our productive and inventive striving (from economic competition to technical innovations). Only in the 20th century did speed begin to receive more systematic, albeit indirect, attention—due to concerns about the unfolding and ultimate effects of widespread acceleration, in terms not only of technical and economic advances but also social change and the overall pace of life.

And when Lewis Mumford, one of the 20th century's most influential historians of technical progress, offered his list of modernity's drivers, he headed it with the power that made higher speeds possible: "Power, speed, motion, standardization, mass production, quantification, regimentation, precision, uniformity, astronomical regularity, control, above all control—these became the passwords of modern society in the new Western style."[18] Or, more programmatically, Virilio's "speed is the hope of the West."[19] But our interest in speed as a remarkable, even magical, and desirable property is ancient, and our quest (often dubious, or outright counterproductive) for higher speeds continues.

The earliest attribution of a flying carpet capable of extraordinary speeds is to a biblical king (Solomon); perhaps its most famous appearance is in *One Thousand and One Nights*, when Husain buys one in India and travels to any near or distant places "in the twinkling of an eye."[20] That—taking for example the distance between Baghdad and Beijing (about 6,300 kilometers), and knowing that an eye-blink lasts about a third of a second—would be about 1,700 times faster than the velocity that large engines impart to rockets to escape the Earth's gravity.

And where would Ulysses be if the goddess Calypso, after detaining him and his crew for seven years, had not then changed her mind and set herself to thinking how she could speed him on his way by providing him with a sharp bronze axe, showing him the largest trees, and bringing him some augers: only that way could he make his escape raft in just four days.[21] And fairy tales, those older and younger than *One Thousand and One Nights*, abound with stories about speed, with characters taking giant steps or running as fast as they can, including a fox (in the Brothers Grimm's "The Golden Bird") that could carry a gardener's son on its tail "over stock and stone so quick that their hair whistled in the wind."[22]

Not surprisingly, modern fairy tales have transformed such achievements into impossibly speedier narratives, in the US perhaps most memorably in the seemingly endless chapters of *Star Wars*. This sci-fi enterprise is replete with transitions of spaceships to "hyperspeed"—travel at speeds faster than the speed of light whose commencement requires precise calculations to avoid collisions with objects that recede around a spaceship's transparent canopy.[23] A less heady, but no less suspenseful, variant of this need for fictional speed are car chases in movies and TV series, and in a multitude of (increasingly realistic) video games. This screen-speed genre is now so common that there are scores of websites ranking the quality of the execution and degree of suspense of cinematic and electronic car (and truck and boat) chases. Some commentators think that speed has become both a cultural and an industrial obsession, a preference in such different realms as arts (film, dance), sports, gaming, and travel.[24]

And seeking faster futures remains an irresistible aim for many businesses. One of the most dubious goals is not only to reintroduce supersonic flight but also to promise that it will become an inexpensive and dominant way of intercontinental travel. Blake Scholl, the CEO of Boom, an American company that claims it will have the first commercial supersonic plane flying by 2027, says that his ultimate goal is to fly "anywhere in the world in four hours for 100 bucks"—although he concedes that getting there might be "two or three generations of aircraft down the line."[25] Indeed, Boom's latest promise of Mach 1.7 by 2029 (jetliners currently cruise at exactly half that speed, Mach 0.85) would be massively short of such a super-speedy goal. "Anywhere in the world" means that an airplane would have to cover up to 20,000 kilometers (the equatorial circumference is 40,000 kilometers) in four hours, and that would require a speed of 5,000 km/h or Mach 4.7, nearly 50 percent faster than the fastest military aircraft (the Lockheed SR-71 Blackbird at Mach 3.2) ever built.

In contrast, quotidian promises of speed have become a common selling point for many kinds of businesses, using that simple marketing claim to differentiate themselves from their competitors. A list of such claims ranges from "fastest from zero to 60" for cars (now featuring new electric models with their swift response) to overnight intercontinental air shipments of packages, and from assorted instant-response 24-hour services (from plumbing to locksmiths) to the now constant promises of the fastest Internet downloads and the fastest communication networks.[26]

All of this means that this book follows a surprisingly under-explored trajectory as it presents a systematic (but still far from comprehensive) inquiry into the manifestations, evolution, histor-ical developments, consequences, and limits of this ubiquitous but (inquiry-wise) strangely sidelined rate. The book's Introduction is dictated by the necessities for proper definitions and well-defined terms—above all making a clear distinction between speed as a fun-damental physical variable (the ratio of distance and time, or speed *sensu stricto*) and its far broader definition as the change of other quan-tities per unit of time (speed *sensu lato*), including (to give just a few disparate examples) the speed of weight gain, learning, computation, GDP growth, and, to note two prominent topics of the 2020s, global warming and the decarbonization of energy supply.

After this, the topical chapters unfold in an orderly, evolutionary manner. First, speeds beyond intuitive human comprehension, from the planet-forming eras and epochs (measured on timescales routinely used only by astronomers and geologists, that is 10^4 to 10^9 years) to the differentiation, growth, and diffusion of life forms evolving on evolutionary timescales. Then, in chapter two, closer looks at speed in the lives of organisms, plants, animals, and humans, from gesta-tion to maturation, from growth to motion, with particular attention paid to two fascinating capabilities: bird flight and human running.

Next (in the third chapter) comes speed in premodern societies, explaining the limits imposed by the time required to produce food by subsistence farming (constrained by the maturation of staple crops and by the speeds with which people and animals can perform repeti-tive, tedious tasks), by unreliable shipping, and by the daily progress of invading armies that was dependent on forced marches, well-fed horses, and, ultimately, on slower draft animals (oxen, asses). The book's penultimate chapter surveys speed in modern societies, from rotation speeds in industrial production, construction, and trans-portation to the enormous speed surge in processing and storing information that has been enabled by the rise and global diffusion of solid-state electronics.

In the closing chapter, I look at the technical, economic, social, political, and environmental consequences of speed, and I offer critical evaluations of speed accomplishments, dreams, and limits, including the claims of the forthcoming singularity. I quantify many notable cases of modern speed accelerations, comparing them to stated aspirations and contrasting them with the boundaries of what is desired and what might be possible: these evaluations uncover the limits to practical speed applications and suggest that a more restrained appraisal of modern accelerations is closer to reality than an uncritical evocation of approaching singularity. A coda of assorted musings brings the book to its speedy conclusion.

My take on speed is guided by my predilection for the unruly realities of the world (rather than for offering dubious grand generalizations). This means that I will offer many generalizations and explain many universal quantitative rules, but I will not ignore the limits of these universalities and will point out some notable departures from such expectations. There is also my propensity for extensive quantification, which is especially relevant here because only hard numbers convey the realities of speed in natural and created environments and inform about the fundamentals, achievements, aspirations, and limits of our lives. And, finally, this apposite reminder: slow, methodical reading will bring greater rewards than zooming through the book at page-turning speed.

Introduction

Universe of Speeds: Variables and measurements

Time and length are two variables describing fundamental physical properties, and their measurements are the foundation on which we have built our understanding of the universe and our planet, as well as the capabilities required to manage increasingly complex economic and social realities. Curiously, measurements of time, the more challenging variable, were regularized long before we got a universal yardstick for length. Our current division of time into 24-hour days goes back to the 2nd century BCE, when Hipparchus of Rhodes (an outstanding astronomer and mathematician, and also the founder of trigonometry) divided the day into 12 hours of light and 12 hours of darkness on the days of the equinoxes.[1]

And even before Hipparchus, Eratosthenes—an Alexandrian mathematician—provided an impressively accurate estimate of the Earth's size (and hence its speed of rotation) by using a simple, ingenious method: at noon on the summer solstice, light cast no shadow as it went right down to a water well in Aswan in Upper Egypt, almost exactly on the Tropic of Cancer, but in Alexandria his gnomon showed a difference of 7° 12'. Given the distance of 5,000 stadia between the two places, he put the Earth's circumference (after adding a small correction to the total) at 252,000 stadia.[2] But because we do not know the exact length of the stadium Eratosthenes used to measure the distance (and because the well is slightly north of the tropic), we cannot pinpoint the actual error: if the stadium was 185 meters, Eratosthenes' result would be about 15 percent longer than the actual length of 40,075 kilometers, and at 176 meters about 10 percent longer. But this uncertainty is much less important than the procedure he followed: calculation based on empirical observation rather than on questionable speculation.[3]

Once the Earth's circumference was known accurately, it was easy to calculate the rotational speed at the equator: it is 1,669 km/h (that is almost exactly twice the cruising speed of jetliners) or about 460 m/s. Studies of ancient and medieval eclipses and the post-1600 observations of lunar occultations of stars indicate that the Earth's rotation has been slowing down, resulting in the lengthening of the mean solar day at an average rate of +1.8 milliseconds per century.[4] But in 2020 the planet had started spinning faster, as that year included the 28 shortest days (by about a millisecond) since 1960.[5] These decelerations and accelerations are far too small to notice.

This long-established calculation of time had no early counterpart in accurate measurement: that came only during the 18th century. After decades of improved designs, in 1772 John Harrison's H5 chronometer—the last in the series of extraordinarily reliable watches he built to win the prize for building a clock whose accuracy could keep a ship within half a degree of longitude after crossing the Atlantic—was off by just a third of a second per day.[6]

Far more accurate ways of timekeeping eventually became available. The old definition of a second as 1/86,400th of a mean solar day (the length of which was fixed by astronomical measurements) was good enough if the precision we required did not go beyond decimal fractions of a second. But in a world suffused with microprocessors whose clock speed is measured in gigahertz—that is, billions of cycles per second—we need a far finer yardstick, and preferably one immune to any changes. We got it in 1967, with the latest revision in 2018—but do not expect that you will immediately grasp the concept.

One second is equal to "the unperturbed ground state hyperfine transition frequency of the caesium 133 atom"—that is, to 9,192,631,770 hertz.[7] Moreover, the atom must be at rest, at a temperature of absolute zero (0 degrees Kelvin), and atomic clocks are also corrected to mean sea level. Other elements could have been chosen, but caesium-133 has two advantages: it has only one stable isotope, which makes it easier to purify (its lighter companions in the group of alkali metals—rubidium and potassium—have, respectively, two and three stable isotopes), and the extent of the cycling of the electron in the atom's outermost shell between two states (after it gets hit by a laser beam) is both large and fast, making its use more accurate than using other atoms.

In contrast, measurements of length remained extremely diverse even well into the 20th century. Well into the early modern era (1500–1800), not only regions and countries but even many towns kept on using different yardsticks.[8]

In December 1799, France became the first country enacting metric measures for length and mass (meter and kilogram), with a meter defined as 1/10,000,000 (a ten-millionth) of the distance between the North Pole and the equator, measured along the meridian running through the Paris Observatory and assuming the Earth deviates from a perfect sphere by a flattening ratio of 1/334.[9] But the adoption of metric units in daily life proceeded slowly even in France. The international unification of measurements began only 75 years later, with the establishment of the General Conference on Weights and Measures (*Conférence générale des poids et mesures*, or CGPM). The CGPM was created by the Meter Convention (*Convention du Mètre*) that was signed in Paris in 1875 by representatives of 17 nations, and held its first conference in 1889.[10] The comprehensive global codification of measurements came only in 1960, with the foundation of the International System of Units (*Système international d'unités*, hence SI).[11]

The entire world is now metric (with the three notable exceptions of the United States, Liberia, and Myanmar); the SI's first base unit of length (*L*), a meter, is now defined with the help of time (*T*) as the fixed numerical value of the distance traveled by light (in a vacuum) in 1/299,792,458 of a second.[12] The speed of light is, obviously, the inverse of that ratio, and it is easier to remember a rounded value of 300,000 km/s. James Kaler, an astronomer, urged us to "have respect for this number: A light beam could travel eight times around the Earth in 1 second and fly from here to the Moon in 1.3. It is the maximum speed allowable anywhere, a fact fully backed up by countless experiments and solid theory."[13]

That ultimate speed also puts immutable limits on reaching any "nearby" star systems and galaxies that might, possibly, harbor other accomplished civilizations. We have identified a number of exoplanets that orbit their stars within what we think are habitable zones (keeping temperatures within limits allowing carbon-based life), but they are hundreds or thousands of light-years away: even if we were able to travel at the speed of light it would take us centuries or millennia to reach them.[14] Of course, speculations about traveling faster than the speed of light and about contacting sapient civilizations will never go away, but the most likely conclusion (given the absence of any clear decipherable signals) is that we are alone and that we have no realistic chance of traveling beyond our solar system.

Speed sensu stricto: *properly defined, observed, and poorly estimated*

Speed, unlike mass or time, is not one of the seven base units of the international system of measurement. Speed is a derived unit, a rate that measures the change of length (*L*, an object's position) per unit of time (*T*) in meters per second (m/s): L/T. That is both correct and simple, but as already noted, Richard Feynman, in his *Lectures on Physics*, called attention to "some rather deep subtleties" that arise when we try to comprehend exactly what we mean by speed, and pointed out ancient Greek confusion about it. Zeno's paradox (Achilles is unable to catch up with a tortoise) is the prime example: the tortoise has a head start, and if Achilles wants to overtake it he must reach at the least the spot where it was when he started running, but by that time it has crawled ahead, and hence he must reach at least that spot, and so on. Feynman points out the fallacy: "although there are an infinite number of steps (in the argument) to the point at which Achilles reaches the tortoise, it doesn't mean that there is an infinite amount of *time*."[15]

A proper definition of speed is to take the ratio of an *infinitesimal distance* and of the corresponding *infinitesimal time*, and then see what happens to it as the time gets smaller and smaller—an option made practically possible by Leibniz and Newton through their invention of differential calculus. As we saw in the Preface, speed is a scalar, a physical quantity that is completely described by its magnitude (as is size or time), while velocity is a vector because it has both magnitude and direction. The quest for calculating the ultimate speed began only during the 17th century, and by the late 1870s we got very close to the actual value.

Determination of universal sizes and distances (radii and orbits of planets in our star system; the diameter of the Sun; distances from our star system to nearby galaxies; the expanse of the universe) and speeds (of the Earth's rotation and orbiting and, above all, of light) was a gradual process extending across more than two millennia, starting in antiquity and culminating in highly accurate measurements done in the late 19th and early 20th centuries. The speed of light remained beyond human reckoning until the early modern era: before that, it was commonly thought to be infinite. The first astronomical (and not very accurate) demonstration of its finite nature was done by a Danish astronomer, Ole Rømer, in 1676.[16]

For years Rømer observed the orbits of Io, one of Jupiter's four satellites, and he realized that the shortening of the interval between successive eclipses when the Earth in its orbit moved closer to Jupiter, and their lengthening as the Earth moved away from that distant planet (repeated observations showed the difference amounted to about 11 minutes), must be due to the time that light needed to reach the Earth. Christiaan Huygens, a Dutch mathematician, then used Rømer's observations to calculate the speed of light as about 210,000 km/s, approximately 30 percent less than its actual value, with the discrepancy explained by Rømer's erroneous estimate of the maximum time delay (the correct value is less than 17 minutes, rather than 22 minutes). In 1849, Hippolyte Fizeau did the first non-astronomical measurement of the speed of light, by sending light through a rotating toothed wheel and reflecting it back with a mirror; by 1879, measurements done by Albert Michelson, an American physicist, came to within 0.02 percent of the correct value; and by 1927 Michelson's new experiments reduced the error to just 0.001 percent.[17]

In 1975, the 15th CGPM (General Conference on Weights and Measures) recommended in its Resolution 2 to use a value of 299,792,458 m/s for the speed of light, and since 1983 this value has also helped to define the length of 1 meter.[18] This advancing accuracy has had many scientific and industrial benefits, but the super-accurate determination of nature's ultimate speed (and indeed of any speed) is not needed in order to be awed by, and afraid of, many ubiquitous natural demonstrations of the *sensu stricto* (*L/T*) speed.

These manifestations range from powerful atmospheric and

geomorphic events (hurricane winds, destructive floods, massive avalanches, and landslides) against which we still have no effective defense,
and which continue to kill and to devastate with unpredictable frequency, to the admirable flight of nimble-footed (deer, antelopes) or
highly streamlined (dolphins, tuna) animals. A book about speed in its
many manifestations must take closer looks not only at these impressive displays of atmospheric phenomena and animal performances in
running, swimming, and flying, but also at the vastly slower but truly
fundamental speeds of terraforming and geomorphic processes.

The creation of continents by the slow but incessant motion
(breaking up, spreading, colliding) of tectonic plates—and their gradual uplift, constant denudation (by glacial, water, wind, and chemical
erosion), and eventual subduction into the Earth's mantle—proceeds
at speeds whose magnitudes are just a tiny fraction of the speeds of
common atmospheric events or the movements which organisms are
capable of. Large continental plates advance just a few centimeters a
year, and their surfaces are eroded by a few centimeters per century,
but these terraforming and geomorphic processes set the stage and
provide the limits of organismic evolution—that is, for differentiation and speciation (the separation of new species).

Unicellular Archean cyanobacteria (their oldest fossilized aggrega-
tions date to about 3.5 billion years ago) remained the Earth's dominant
life forms for nearly 3 billion years before the evolution of the first
highly differentiated multicellular macroscopic animals began more
than 600 million years ago. These developments are traced in the first
chapter. Early organismic evolution proceeded at speeds even slower
than those of tectonic and geomorphic processes, but the subsequent
rates of increase in the typical length, volume, and mass of life forms
were much faster, culminating in the appearance of the largest terrestrial animals—the Cretaceous sauropods (long-necked, long-tailed
dinosaurs) that walked the Earth between 145 and 65 million years ago.

The resulting speciation led to three dominant modes of
locomotion—on land, in water, and in the air—each with some distinct modifications: walking and running, but also slithering and
jumping, on land; flight by flapping wings, but also by gliding and
soaring on rising warm air; and from the flagellar swimming of

microorganisms to pulsating jets of octopi and tail-powered swimming of fishes in water. All modes of organismic locomotion will be considered, and their speeds (and their consequences for survival) will be explained in the second chapter, before I turn to locomotion in humans. While we cannot fly (by flapping or gliding, although accomplished cyclists can power lightweight flying machines by pedaling), we are excellent walkers and runners and reasonably good swimmers. Again, I will look not only at the common and extreme speeds of our locomotion, but also their impacts on our evolution and well-being.

Traditional human societies were energized by human and animal muscles (that is, ultimately, by the conversion of solar radiation to edible biomass), by the burning of wood (or charcoal derived from it), and by wind and flowing water converted by mills and sails. These animate prime movers, simple machines, and (mostly wooden) devices available in premodern societies created the world of slow to moderate *sensu stricto* speeds. The limited range of these speeds on land had a span of just one order of magnitude, from the slow uphill walking of heavily laden porters in mountains (less than 4 km/h) to riding a galloping horse (a typical speed of just above 40 km/h)— but while forced army marches, even carrying full loads, could be endured for many hours, a horse's gallop (the fastest mode of equine locomotion) would last only a few kilometers without interruption.

And a much narrower range prevailed on water, as people paddled canoes or kayaks (3–5 km/h) or manned sailships running downwind: in the ancient world, ships with simple square sails could rarely cover more than 7 km/h, and even by the 19th century the record times posted by fully rigged clippers (whose speed on long-distance runs would remain unsurpassed by any sailing vessel for a century) translated to no more than about 25 km/h when averaged over the entire course of their intercontinental journeys.[19]

Obviously, these restricted speeds on land and water determined the rapidity, frequency, and intensity of long-distance trade, as well as the progress of maritime exploration and the daily advances of large invading land armies whose maintenance required a great amount of supporting materiel and food transport that could be moved only by slow ox trains.[20]

Speed became a defining feature of modernity. During the 19th and 20th centuries, many technical advances were driven explicitly by the quest for higher speeds. Between the 1830s and 1900, new prime movers, new materials, and new designs set a series of speed records in sailing, long-distance steam-powered shipping (with the North Atlantic crossing and the UK–Australia run as the two key measures of progress), and intercity travel by trains (speeds of 100 km/h had already been reached very briefly before 1850). After millennia of road speeds constrained by the limited powers of animal draft, travel on land began to be transformed by internal combustion engines in cars whose maximum speed reached 100 km/h in 1902, just 17 years after Benz built his first small, slow, carriage-like, gasoline-fueled Patent-Motorwagen in 1885.[21]

The admirable speed of inventive engineering: 17 years from Benz's rickety motorized three-wheel carriage to an essentially modern automobile, the 1902 Mercedes-Benz.

The first half of the 20th century was marked by steady improvements in the speed of flying. The first slow, brief, ground-hugging flights of a machine heavier than air were made by Wilbur and Orville Wright above the Kitty Hawk dunes of North Carolina on December 17, 1903. The longest flight of that day, which wrecked the canvas-and-wood plane, was 260 meters in 59 seconds (4.4 m/s), while the cruising speed of a Boeing 707, the first jetliner in transatlantic service, was 896 km/h, or 250 m/s, a more than 50-fold gain in just 55 years. The fastest sailing ships had increased their speeds only four- to five-fold during the 350 years between 1500 (caravels) and 1850 (clippers). The first scheduled jetliner took off in May 1952 (but the de Havilland Comet's accidents ended its first-generation service just two years later); regular transatlantic flight commenced in 1958, and the first wide-body jetliner, the Boeing 747-100, was first flown in 1969.[22]

An even more impressive speed jump: from the Wright brothers in 1903 to the first (transonic) Boeing 747 in 1969.

The concurrent rise of all speeds of mechanized travel—including ocean-borne shipping, using larger bulk carrier and container vessels, and rapid electric intercity trains, with the first service starting in 1964 in Japan—changed many economic arrangements and resulted in some widespread organizational shifts. Certainly, the most notable examples of these changes have been the globalization of energy supply (the shipping of coal, crude oil, and liquefied natural gas), materials and food supply (bulk carriers moving ores, cement, fertilizers, grains), and an unprecedented dependence on imported manufactures (with specialized ships and cargo planes carrying everything from cars to toys and from clothes to electronic gadgets).

Reliable speed has also enabled just-in-time deliveries for assembly lines (eliminating the need to maintain large, expensive inventories) and resulted in many behavioral changes and quality-of-life improvements (including previously impractical daily commutes to work, and affordable family vacations on distant continents). At the same time, it has also generated unrealistic expectations about the further, continuing progress of travel speeds. Calls for the mass-scale construction of near-sonic travel corridors on land (magnetic levitation and "hyperloop" trains) and renewed attempts to revive rightly unsuccessful supersonic flight have been the two most recent prominent examples that will be assessed later in this book.[23]

All speeds can be accurately measured in a variety of ways. Onboard meters in vehicles, trains, ships, or planes give very accurate readings—as does radar (reflected electromagnetic waves), used to enforce road speed limits, follow jetliner flights, or monitor bird migration. Small, light GPS devices fastened to animals, birds, and even insects can provide reliable extended measurements of speeds deployed in the daily lives of organisms. But without meters, reflected waves, or careful counting, we are not very good at making accurate unaided observations of speed, and in the absence of additional sensory stimuli we do not perceive constant speeds. A high-speed train covers more than 80 m/s (nearly 300 km/h), a jetliner cruises at 250 m/s (900 km/h), the Earth at the equator is rotating at 460 m/s (1,669 km/h), and the planet travels nearly 30,000 m/s (107,000 km/h) on its elliptical orbit around the Sun—but we do not have the slightest perception of the two planetary motions, and with closed eyes, or with shades lowered, we cannot provide any perception-based estimate of speed for the first two experiences. That changes only when we register rapidly changing surroundings.

James Joyce, in *A Portrait of the Artist as a Young Man*, describes how the protagonist "saw the darkening land slipping away past him, the silent telegraph-poles passing his window swiftly every four seconds"—and even made the words of his silent prayer "fit the insistent rhythm of the train; and silently, at intervals of four seconds, the telegraph-poles held the galloping notes of the music between punctual bars."[24] If the traveler knew the standard distance between those poles (60 yards or 54.9 meters), he could easily calculate the train's speed: just short of 50 km/h, a rather typical performance for a late 19th-century train, and one not inducing any sensory discomfort.

But the passing-poles experience may be annoying when masts placed at 50-meter intervals to support catenary wires for high-speed electric trains swish by in less than a second. And some people feel discomfort and even experience motion sickness when facing rearward, as they observe rapidly receding nearby features seen from their window. This is not a problem on any of Japan's Shinkansen trains, where all seats can be rotated (by the staff at terminal stations) 180 degrees so they always face forward (though if a small group wants to sit facing each other, passengers can turn some of them back). And facing rearward on an intercontinental flight (some airliners, including British and Qatar Airways, have such seats in business class) makes no difference compared to the dominant frontward-facing seating when it comes to seeing the same cloudless sky, or just a thick cloud layer or the night's blackness as the plane cruises at more than 10 kilometers above sea level.

All of us react only to sudden changes of speed, to deliberate or unexpected accelerations or decelerations. You feel such accelerations every time a plane is taking off (as your body gets pressed into the seat and as any loose objects slide or fly backwards), and even more so during a go-around—an aborted landing—when a pilot or air traffic controller decides that it is unsafe to land and the engines accelerate the plane as it rises again, to make a circle or an ellipse before positioning for another approach. My most memorable go-around was in the late 1990s at Heathrow when, after a transatlantic flight, our Boeing 747 had just flown above the white threshold stripes marking the runway's beginning when the pilot pitched the nose up and the four large accelerating engines lifted the jumbo plane once again high into the air: such a speed surge, executed so close to the ground, was as impressive as it was frightening.

As for everyday experiences, there is now a rich experimental literature on speed perception that demonstrates how assorted variables affect the subjective experiences of fast movements. For example, car drivers and cyclists sharing urban roads have different perceptions of speed and relative safety depending on street configurations (narrow, wide, twisting, sloping), and car drivers tend to underestimate actual speed as the field of view increases and overestimate speeds slower than 50 km/h.[25, 26] More importantly, advertisements encouraging drivers to slow down may not lead to desirable outcomes:

> Those who believe they drive faster than average will accept the advertising message as aimed at themselves; those that believe they drive slower than average will view the campaign as aimed at others. The problem remains that a substantial number of drivers falsely believe they drive slower than the average driver, and drivers generally exaggerate the usual speeds of others.[27]

Pedestrians make equally unreliable judges of speed when crossing an unmarked highway. In experiments with a single vehicle, participants underestimated higher (more than 50 km/h) speeds; when two approaching vehicles were involved, they underestimated lower speeds (less than 50 km/h).[28] And then there is a well-established phenomenon of the perceived speed of moving objects depending on image contrast.[29] Lowering the contrast decreases perceived speed, but this effect weakens at higher speeds.[30]

Finally, a few comments about wind speed, a commonly encountered continuum that needs some classification. In 1805, Francis Beaufort (an Irish hydrographer, and eventually a British rear admiral) proposed a scale for the speed of wind, and it was used for the first time by Captain Robert FitzRoy—who commanded HMS *Beagle*

BEAUFORT WIND FORCE SCALE

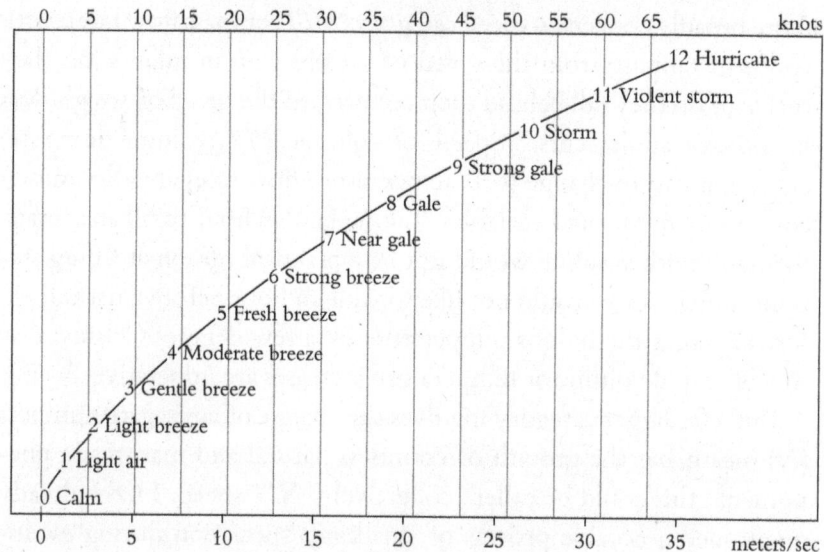

From calm to hurricanes: the Beaufort wind force scale.

on its circumnavigation of the Earth (1831–1836) with young Charles Darwin on board.[31] The scale divided wind speeds into 12 force categories, from calm (<1 km/h) to hurricane (>118 km/h), and offered helpful perception hints to separate the adjacent categories. Light breeze (6–11 km/h): "wind felt on face; leaves rustle"; gentle breeze (12–19 km/h): "leaves and small twigs in constant motion . . .; light flags extended . . ."; moderate breeze (20–28 km/h): "raises dust and loose paper; small branches moved . . ."; fresh breeze (29–38 km/h): "small trees in leaf begin to sway; crested wavelets form on inland waters . . ."

Speed sensu lato: *beyond the strict physical definition*

Speed, a time-limited rate, must always have time as its denominator, but we have been using the noun to describe many phenomena that do not have length (distance) in the numerator. These *sensu lato* measures of speed count the rate of change of other basic physical entities (mass, volume, temperature). Mass (M) is a common numerator within this more broadly conceived speed category (M/T, or mass flow rate), with concerns ranging from the speed of weight gain in infants (do they thrive or do they fall behind their cohort?) to the speed of weight loss in anorexic adolescents.[32] Speeds of volume (V/T, volume flow rate) and temperature change (t/T, temperature flow rate) are also among commonly monitored variables. Taking half an hour to fill an average fuel tank with gasoline would not be a practical option at filling stations, much as we would not like to wait an hour before a natural gas furnace raises the indoor temperature by a few degrees Celsius: reasonably rapid volume or temperature increases are imperative.[33]

But a far larger category involves the change of aggregate numbers (N) measuring the growth of countless natural and man-made phenomena: this could be called, collectively, N/T speed. I have already mentioned a notable process of this kind: speciation during evolution. How long did it take for a group within a species to separate itself from other members and to develop unique characteristics and reproductive incompatibility? How many new species arose during specific periods of time? Were there some remarkably fast and some inexplicably slow periods of speciation, besides the famously prolific great Cambrian diversification that began about 540 million years ago? Looking at recorded history, we might do similar comparisons for the speeds with which new empires formed, endured, and unraveled.

In modern societies, this N/T speed usage is ubiquitous: "how long does it take" and "how fast can it be done" are among the most common concerns and preoccupations of the modern age. We often label such phenomena with the cognate adjective (speedy service, reply, return), with its synonyms (fast, rapid, swift), or with other qualifying adjectives (record-breaking, dizzying, unexpected, unprecedented, soaring). Recent examples of notably speedy phenomena range from China's fast post-1990 economic growth—sometimes even in the double digits per year, resulting in the quadrupling of GDP between the years 2000 and 2023[34]—and the unprecedented global market penetration by mobile phones, from 1 percent of the world population owning a device in 1994 to 50 percent by 2007 and 85 percent by 2023.[35]

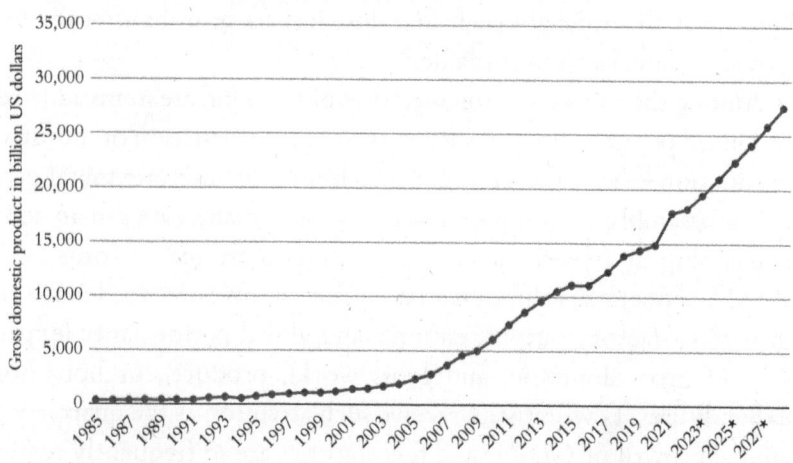

Gross domestic product (GDP) at current prices in China from 1985 to 2022, with forecasts until 2028*

Source: IMF © Statista 2024

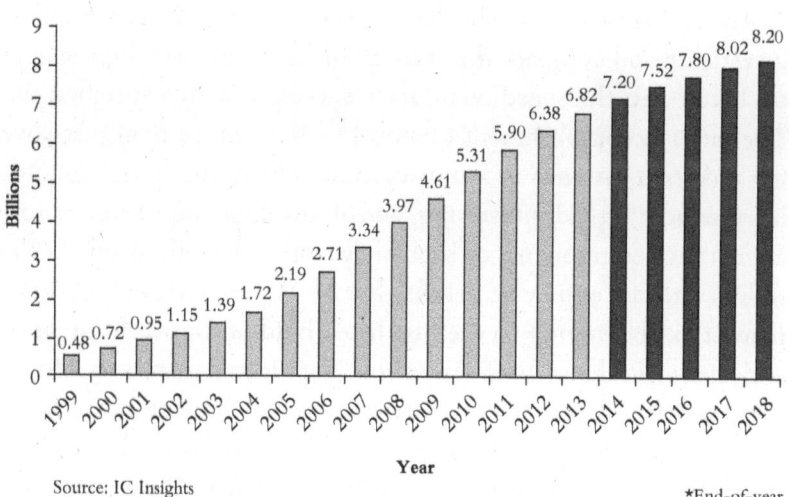

Total worldwide cellular subscriptions*
(1999–2018)

Source: IC Insights   *End-of-year

China's rapid GDP growth and the global adoption of mobile phones are two much-quoted examples of impressive N/T speed.

Another prominent example of a much sought-after and eagerly expected speedy action was due to the Covid-19 pandemic: the World Health Organization proclaimed its beginning on March 11,

2020; the first vaccines became available in the US before the end of the year; and in Canada, a country with a high vaccination rate, 65 percent of all adults got their first dose just six months after the first Covid vaccine became available.[36]

Among the most common measures of this kind are items added or produced per unit of time, such as speed of construction or industrial production—be it the precision-machining of intricate metal parts or the assembly of complex machines (how many cars can an automaker ship in a year?). Even more common are money/time ($/T$) speeds: monetary values created, spent, or invested to denote the growth of factory output, national and global performance (expansion of gross domestic, and gross world, product), or household expenditures. Few statistics receive such attention as the quarterly or annual growth of GDP—and few statistics are so frequently revised just a few months or a year later, to say nothing about their multiple inadequacies to capture an economy's real progress (that is, in terms of improving everyday lives and reducing income inequalities).[37]

And as no historical era has been without its armed conflicts, we can investigate and compare the establishment of military supremacy—or, better yet, the speed of military success within a specified time. The latter accomplishment obviously entails more than just covering a certain distance in a unit of time during the initial conquest, because holding, administering, and absorbing conquered territories are decisive components of such endeavors: the high speed (L/T) of military advances may be helpful, but N/T speed (with N subsuming many tasks belonging to the just-listed hold-and-govern categories) is decisive. Perhaps most famously, Napoleon's march to Moscow was as swift as horse- and ox-drawn vehicles allowed, but it had no hold-and-govern sequel, just a swift, ignominious, and devastating retreat: of 422,000 troops that crossed the Neman River going east, only 10,000 men returned.[38]

Few recent endeavors have illustrated this difference as starkly as the US involvement in Afghanistan: that neutral, descriptive noun (involvement) was chosen deliberately, as it was neither a real invasion and full-scale war nor a real occupation and a resolute quest for subjugation. L/T was (militarily speaking) unexpectedly swift:

the Taliban left Kabul on November 13, 2001, less than two months after the first CIA officers were flown into the Panjshir Valley (on September 26, 2001), and within a few months the US military established its fortified bases in support of a new Afghani administration throughout the country.[39] But neither the US nor the series of subsequent Afghani governments and their hollow army managed to take dominant political and social control of the country during the next two decades, and the return of the Taliban in August 2021 was even swifter than its retreat, as the entire country fell under its control in a matter of hours.[40]

But perhaps the most common of all *sensu lato N/T* rates has been one of the key derived scientific units: power is defined as energy (received, extracted, produced, converted, expended) per unit of time, with one watt (W) equal to one joule (J) per second (W = J/s). But a joule itself is a derived unit, so it all gets a bit more complicated. The International System of Units defines a joule as the work done by a force of 1 Newton (N) moving 1 meter; and, in turn, 1 Newton is the force required to accelerate the mass of 1 kilogram at a rate of 1 meter per second squared, making 1 joule equal to 1 Newton-meter.

But originally, long before the SI derivations, a unit of power was defined in an empirical way by using a conservative estimate of the typical speed of a later 18th-century English horse harnessed to work by walking in a circle. There have been various explanations of the unit's origins, but the only correct one can be traced to an entry James Watt made in his *Blotting and Calculating Book* in August 1782, when he needed to calculate the power of his steam engine that could replace 12 horses working in a paper mill.[41] He did not carry out experiments but referred to Mr. Wriggley, the millwright, who told him that a mill-horse walking in a circle with a 24-foot diameter makes two and a half turns a minute. That is a speed of 3.14 feet (0.96 meters) per second and 3.45 km/h—slower than the typical speed of people walking slowly without any load (4 km/h). But modern measurements of working horses confirm that value, as they put typical speeds at 0.9–1.1 m/s.

The distance covered by a mill-horse in a minute was thus 188.5 feet ($2\pi \times 12 \times 2.5$), but as Watt used 3 (rather than 3.141) to approximate π,

Harnessed horses walking in a circle: how horsepower was defined by James Watt
with the help of a millwright friend.

his result was just 180 feet. The experienced millwright also estimated
that his animal worked at the rate of 180 lb per horse, and hence the
work calculated by Watt was 32,400 foot-pounds (ft·lb) rather than
the correct value of 32,900 ft·lb per minute. In September 1783, Watt
settled on a more accurate approximation of 33,000 ft·lb, and this
unit—1 horsepower (hp)—became the standard measure of power in
British engineering practice and in the International System of Units,
where the unit of power is named in Watt's honor: 1 horsepower is
equal to 745.7 watts.

Watt worked with good numbers, but his value for 1 horsepower
was higher than could be expected from most horses harnessed for
work during the 18th century (and for many preceding centuries).
Some heavier, well-fed animals could sustain power close to 750
watts for hours, but smaller and inadequately fed animals delivered
no more than 0.6–0.7 horsepower. Watt may have been aware of this,
and by choosing a relatively high value he may have avoided any
disappointments about the performance of his engines: if an owner
replaced three horses with a 3-horsepower engine (working tirelessly
at that full rate for as long as a boiler was stoked sufficiently), he
would have had plenty of power to spare.

Power is thus an indirect measure of speed: everything else being equal, the more power we deploy, the faster we finish a task. To plow a hectare (100 meters × 100 meters) of land before planting spring wheat, a 19th-century Dutch farmer working with two good horses (a combined power of 1,200 watts) needed half the time it took a Roman farmer who worked with two oxen (a combined power of 600 watts) at the beginning of the common era—and he did so despite heavier soil, because he also had a better, metal plowshare.[42] In 2020, a large American tractor (300 horsepower, or about 225 kilowatts—that is, nearly 200 times more powerful than the Dutch horse team) could finish plowing a hectare in less than 10 minutes.[43]

Similar speed increases apply to all other field operations, and these huge differentials made possible by the mechanization of farming are the principal reason for the decline of the agricultural labor force from the dominant category 200 years ago to the lowest share among all major economic activities in affluent countries of the early 21st century.

And similar historical trajectories and similar implications—of tasks completed faster; of human (or animal) exertion lessened or entirely eliminated; of time saved that could be spent on other activities (and also of achieving previously unthinkable feats)—hold for all energy converters, ranging from diesel-powered water pumps used for irrigating fields (a yield-increasing activity that began millennia ago with the laborious manual lifting of water pails from wells or streams), to electric motors used to blow a strong stream of air-conditioned air (contrast that with the tiresome work of men or boys using pulleys to operate a large swinging fan in pre–air conditioning tropical regions). Some of these contrasts have been stunning: during the 1950s, jet engines (gas turbines) powering modern airliners made the cruising speed of modern airliners about 30 percent faster than the fastest speed of a P-51 Mustang or a Supermarine Spitfire, two famous Second World War fighter planes.[44]

In the world beyond the strictest physical definition, all measures of power as well as all measures of growth that have time in the denominator are thus measures of speed. That qualification for growth is necessary, because we can measure growth with other variables in the denominator. A very useful (and already noted) example is the weights of children vs. their heights: the graphs (or tables) enable us to see instantly if a child is underweight, has an ideal weight, or if they are overweight or obese. An entirely different but no less useful example relates monetary outputs to energy inputs in economies: obviously those with higher GDP/energy ratios will be more efficient users of fuels and electricity, generating lower environmental impacts. But caution is needed: countries may use less energy per unit of GDP because (as most Western economies have done) they have offshored many energy-intensive industrial tasks to lower-income countries in general and to East Asia and China in particular.[45]

In traditional societies, all L/T speeds of human activities were low, and so were all kinds of N/T (or $\$/T$) speeds: indeed, in some traditional societies most of the changes that could be measured by *sensu lato* speeds were entirely or nearly absent. Perhaps most notably, global population growth remained stagnant or very low for more than 90 percent of recorded history.[46] This low rate of population change both reflected and affected the centuries of very slow economic development. During antiquity and the Middle Ages, the world could not be preoccupied, as we are today, with the growth of economic product: even if it could have been, somehow, measured (a challenge in societies where so many transactions were yet to be monetized), the numbers would have shown hardly any year-to-year change.

We know that for certain because economic historians have devoted a great deal of effort to reconstructing past rates of economic growth for those societies where various preserved partial records of material output, wages, prices, and consumption have made it possible to calculate long-term trends with satisfactory approximation. Their results, including some long-term reconstructions of staple crop yields, will be described in some detail in the third chapter of this book, and they have convincingly indicated either no long-term growth or minuscule annual gains, even in regions that were organizationally and technically rather advanced.

This all began to change, slowly, during the first two centuries of the early modern era (1500–1800), and these incipient moves toward faster population growth (the global rate doubled during the

17th century), wider urbanization, industrialization, globalization, and rising private consumption (driven by a combination of factors ranging from scientific and engineering advances to European empire-building) rapidly intensified after 1850. By 1914, when the First World War began in Europe, the lives of average citizens in the world's two economically most advanced regions (Western Europe; the US and Canada) were, by all reckonings, further away from the lives of their predecessors who lived in 1814, the year of Napoleon's exile to Elba, than the lives of peasants who worked the grain fields of Europe and Asia in 1814 were from the quotidian experiences of their forebears who lived in 1514, three centuries before them—at the very outset of European colonization forays (Vasco da Gama reached India in 1498).[47]

And, despite a protracted economic crisis in the 1930s followed by the even more devastating Second World War, the speed of population, technical, and economic gains further accelerated after 1920, and the 20th century offered an unprecedented range of speed gains. The rising speeds of industrial output (resulting from mechanization and automation) and of communication and information (resulting from the adoption of solid-state electronics) have been among the most impressive entries in this large category. The number of products—from aluminum cans to water pumps—coming from automated factory lines per hour or per year measures the speed of modern industrial production, with high rates conferring a competitive advantage; while words typed per minute and a computer's processor clock speed (how quickly the central processing unit can retrieve and interpret instructions given to it) are common examples of measuring the speed of processing transactions.

Examples abound. A modern blow-moulding machine will make up to 12,000 PET (polyethylene terephthalate) beverage bottles per hour compared to 200 same-sized glass bottles made in 1900 by a semi-automatic machine. By 2020, Toyota, the world's leading carmaker, produced about 9 million vehicles a year—about 3,000 times more than in 1945, the year of Japan's defeat.[48]

While an experienced typist could do 80 words per minute (twice the average non-professional speed) a century ago or today, the clock speed of personal computers (measured in hertz, or cycles per second) is now three orders of magnitude (more than 5,000 times) faster than in 1971, when Intel released its first microprocessor: 4.2 gigahertz (GHz) vs. 740 kilohertz (kHz).[49]

Yet another common usage of speed refers to many situations where not only is rapid execution desirable, but the duration of an intervention has life-and-death consequences: the speed of mobilization in a country under attack; the speed with which a government delivers essential aid to a region stricken by a major natural disaster (almost always found to be too slow by the would-be recipients); the speed with which a stroke victim is treated in hospital (the fastest possible intervention is tied to the best possible outcome).[50] In all of these cases, time is always the only denominator, but the numerators are more or less complex assemblies of variables that include not only distance—in the case of making Toyota's cars, those considerations range from the timely imports of raw materials for producing steel and the smooth functioning of long, continuous, just-in-time assembly lines, to the logistics of exports reaching all continents—but also the availability and deployment of appropriate expertise, machinery, and, if need be, capacity for crisis management.

These variables can be used to set desired speeds of production, processing, or problem resolution, and in all such cases speed is an excellent proxy for gauging an organization's, or a nation's, capability to manage complex tasks and to stay competitive in the fast-changing global market. Perhaps the best, all-encompassing measure of this ability has been quantified by the economic complexity index (ECI), which measures both the diversity of a country's exports and their sophistication.[51] Countries with a low ECI export only a limited number of widely available products (such as crude oil or apparel), while highly diversified economies sell goods and services of low ubiquity that require advanced production techniques (microprocessors, pharmaceuticals, advanced medical equipment).

Few measures offer a better insight into the changing fortunes of nations than looking at the direction and the speed with which national ECI rankings change. Between 1995 and 2020, China rose from 46th to 17th place—but that rise was surpassed by the speedier

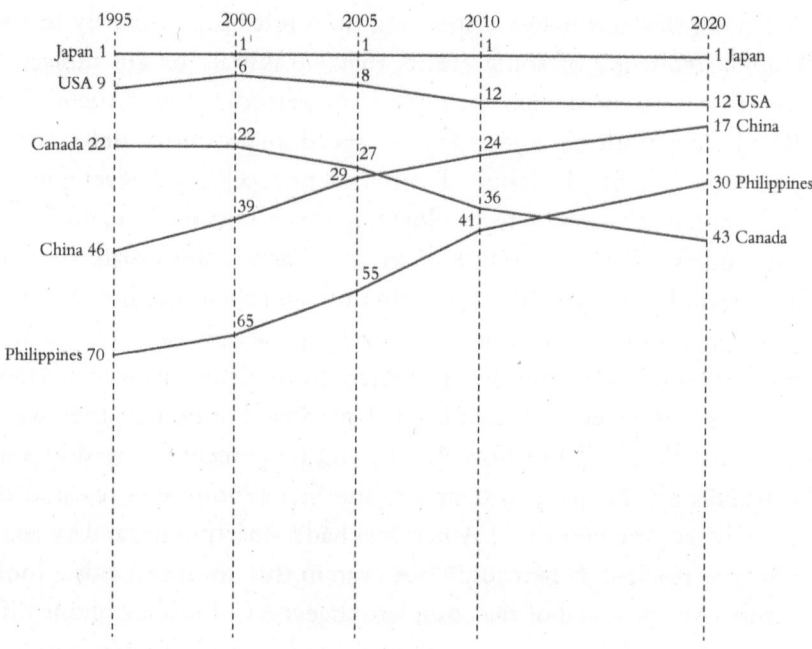

National indexes of economic complexity have shown remarkable stability (Japan), impressively rapid gains (China, the Philippines), and regrettable decline (Canada).

elevations of Vietnam (from 107 to 52) and the Philippines (from 70 to 30); and while the US ECI declined only slightly from 9th to 12th, Canada's performance plunged from 22nd to 43rd place, well below the Philippines as well as Turkey, Panama, Bulgaria, and Bosnia and Herzegovina. Clearly, in light of this important measure, Canada does not belong among the G7, not even among the G20, and the speed of this economic deterioration has been depressing.[52]

Finally, the last addition to the large class of *sensu lato* speeds is the one referring to the accelerating pace of life—to the general speeding-up of human affairs. Time on the Earth cannot go faster (there is no speed of time), and so the expression "time flies faster" can refer only to our perception of it. Experimental studies have confirmed that this perceived (subjective) "speed of time" varies among individuals, and that it is influenced by mental and physical state as well as by age: the older we get, the more we perceive time going faster.[53]

Saying that our lives are speeding up is referring primarily to the higher frequency of some events that we initiate or are subjected to—that is, to the shrinking interval (the period) between them. The Rand Corporation's 2018 study on speed and security defines this reality as "a heuristic designed to represent accelerated development and implementation of technologies across a host of domains," and the study's opening sentence claims that "across the world, there is a profound sense that life is speeding up, and the faster life gets, the more we have to adjust our norms to keep pace."[54] Communication is the most obvious, and the most cited, example of this acceleration.

Telephones were invented in the late 1870s, but by 1946 they were only in half of all US households, and 90 percent ownership was reached only in 1975; in contrast, the first iPhone was released in 2007, by 2010 20 percent of Americans had a smartphone, and by 2022 that rate reached 97 percent.[55] But even in this instance a closer look shows that the speed of the complete trajectory of mobile phone diffusion in the US was not so unprecedented, as it was nearly matched by the adoption rates of radio, color TV, and the microwave oven. In 1983, the DynaTAC 8000X became the first commercially available cellular phone, and 90 percent ownership was reached 20 years later—while US radio ownership, despite the economic crisis of the 1930s and the Second World War, went from zero to 90 percent of all families in 22 years (1924–1946), color TV reached 90 percent of households after 24 years (1960–1984), and the ownership of microwave ovens rose to 80 percent of households just 17 years after the gadget's introduction in 1967.[56]

But these comparisons ignore the different modes (qualities, ranges) of use—the distinction that is most obvious when looking at a landline phone and a smartphone. Landline phones carried only voice and were picked up only when they rang or to make a call, which means that many people talked for hours every day (unlike in Europe, flat rates for North American intracity calls made that costless)—while other people did not pick up their phones for days. The best available data (from the US Bureau of Labor Statistics) show that in 1990 (before mobiles came into a wide use) an average American spent about 12 minutes per day talking on a landline. In contrast, smartphones make it possible to receive an entire universe of still and moving images, and they are used repeatedly for sending anything from weirdly worded brief messages to videos. Even so, the results of a 2022 survey of American phone habits by Asurion, a phone insurance and tech care company, were astounding: adults above 18 years of age checked their mobile an average of 352 times per day (that is, every 2 minutes and 43 seconds during waking hours)—and that frequency was nearly four times as high as in 2019.[57]

Moreover, 64 percent of Americans use their mobiles while on the toilet, and 45 percent say that the gadget is their most valuable possession. No wonder that 45 percent also admitted that they are mobile phone addicts. But the most stunning of all is this fact: in 2022, an average American spent 2 hours and 54 minutes on a mobile phone every day (that is 44 days a year), and even higher figures were reported for 2023—4 hours and 30 minutes according to Backlinko (a search engine optimization website) and 5 hours 24 minutes according to Vecta Labs (a telecommunications service provider).[58]

At the same time, surveys and statistics show other increased frequencies: taking shorter vacations more often; eating out more frequently; getting retrained or updated more often as software and operating procedures keep changing.[59] Add to this the changes in everyday actions, from the now ubiquitous online banking to going through automated passport control: the norms that prevailed for decades change suddenly, and then change again, as machines and electronics keep displacing people.

And, increasingly, experiences of these higher frequencies, previously limited to affluent countries, have become common even in many low-income societies, ranging from near-instant price information and rapid trading via mobiles in African food markets, to rapidly shared tips and notifications about the best ways of illegal border crossing. There is also no doubt that the increased frequency of some changes is already making our decisions more difficult (and more consequential), and that the higher pace of technical advances will demand even faster reactions when encountering such accelerating shifts as the widespread and poorly regulated deployment of artificial intelligence, resulting in the mass dissemination of fake information.

Such fears have already led to many speculations (from sensible to dubious, from plausible to rank sci-fi): daily lives getting "faster" in so many ways that we will simply lose control unless we can, somehow, shift into "hyperdrive" to cope with a world dominated by ubiquitous AI, rampant genetic engineering, the 3-D printing of body parts, cognitive brain implants, near-instant intercity travel, and interplanetary flights, and threatened by hypersonic missiles and the mass-scale denial of all electronic access. Some of these wonders or perils are promised to be as ordinary (or as likely to happen) by 2040 as mobile phones are today![60] Concerns about their impacts are, undoubtedly, justified, but in this book I will concentrate on narrowly defined speed (L/T) and on the principal categories of N/T—output (materials, finished products), power inputs, and population and economic growth—rather than on the increased frequencies of events, habits, and choices, variables that have a considerable qualitative component and are not as easily captured by numbers.

This, then, is the universe of speeds, from intergalactic light to shifts in economic complexity, from fleeing antelopes to soaring albatrosses, from intercontinental flight and shipping to the economic fortunes of countries and empires, from barely noticeable advances to concerns about accelerated change surpassing our abilities to cope. The remit is, all too obviously, far beyond any detailed and comprehensive coverage. There are many ways to deal with this challenge, and I have chosen a combination of basic (and generic) explanations of all essential speed phenomena (what determines the speed of running or flying; what innovations made it possible to multiply the speed of computing) and a closer look at some prominent examples in every major speed category.

That is how wind erosion, hunting cheetahs, galloping horses, elite sprinters, clipper ships, giant container vessels, and Boeing 747s get special attention in the narrower L/T category, and how the power of prime movers, the speed of metal machining, America's wartime industrial mobilization, Toyota's Kaizen production system, the decades-long run of Moore's law, and the likely speed of global energy decarbonization are featured in the broader N/T category. Presenting this wide-ranging analysis of speeds fills a surprising void in the modern coverage of this fundamental variable, and as always in my attempts to deal with complex realities, history will have an inevitably prominent place in my explanations because its understanding helps us to appreciate the preconditions, circumstances, limits, and possibilities of the unfolding events. And, once again, this book is full of numbers: this relentlessly quantitative approach may not be popular, but it is imperative for a proper understanding of the world and its dominant speeds. There is no better antidote to wishful thinking and fact-free judgments and forecasts.

1. Evolution

Non-intuitive time spans and speeds

There are no easy, intuitive ways to grasp, to distinguish, to compare, and to truly comprehend very large numbers—and I do not mean such mind-defeating ones as a googol (10^{100}—ten raised to the power of 100) or googolplex (10^{googol}—10 raised to the power of a googol).[1] I am talking only about rather mundane totals of tens of thousands (10^4), hundreds of thousands (10^5), millions (10^6), tens of millions (10^7), hundreds of millions (10^8), and billions (10^9) of years. Mundane because we encounter them repeatedly when describing our star system, our planet's origins, the formation of its oceans and continents, the emergence and evolution of life, and the history of our species. The mental challenges of large numbers are well known, because throughout their evolution our brains have been optimized for dealing directly with a small number of items (other individuals, tools, animals).

As quantities increase, we must count the increments, but very large numbers are beyond any practical counting: they become categories rather than single entities. While a true, intuitive comprehension may be unattainable, you can train yourself to think more like a geologist, an evolutionary biologist, or a paleoanthropologist, and to appreciate more fully the lengths of time (and hence the speed of transformation) required first to form a planet habitable by our—carbon-based—form of life, then the differentiation of these life forms into millions of species. And then the time needed for the emergence of an organism sapient enough to superimpose another envelope—that of the technosphere—on the planet's unique biosphere, the some 20-kilometer-thin layer of life (compared to the Earth's radius of 6,371 kilometers) that encompasses the planet's atmosphere, the entire hydrosphere (living organisms can be found in the deepest ocean trenches, 11 kilometers below sea level), and the uppermost strata of the Earth's lithosphere (bacteria have been found in the deepest mines).[2]

The aim of this chapter is not to provide yet another chronological account of the Earth's history, reviewing first the sequence of geological eras and epochs and then the progress of life's evolution; there is no shortage of such grand overviews. Instead, I will note the most notable milestones in the Earth's physical evolution—from the planet's formation to the appearance of its oxygenated atmosphere—in order to focus on the durations of these events (their speeds), and then to emphasize the accelerating development (of course, the qualifier is relative) thanks to which living organisms got more complex and adapted so successfully that they now inhabit every niche within the biosphere.

The long times required for forming, and transforming, the Earth's large-scale physical features—its oceans and continents, and its diverse surfaces ranging from tall mountain chains to extensive high plateaus or low-lying plains—mean that we are dealing with speeds that are also outside our quotidian experiences, but in reverse. In these instances, we are not overwhelmed by large magnitudes but must reckon with speeds that are far below anything we encounter in our daily life. Compared to common speeds of walking and running (4–10 km/h), or driving (40–120 km/h), tectonic and geomorphic changes (the creation and destruction of continents; the formation of surface features) proceed at speeds that are many orders of magnitude slower, although some of their agents, from volcanic eruptions to hurricanes, do unfold at very high, albeit time-limited, speeds.

The creation of this new Earth's crust proceeds at speeds of no more than about 20 cm/year—seven orders of magnitude slower than slow walking (conversely, walking is about 175 million times faster).[3] And while we can relate to speeds of water and air, the two key agents of erosion and denudation (raindrops falling, streams flowing, and winds blowing are within the compass of our repeated experiences), their cumulative actions are, except in the case of such overtly catastrophic events as enormous floods or storm surges that rip away riverbanks or ocean shorelines, beyond our range of near-term perception. Steady cumulative erosion and denudation wears away hard rocks by as little as a few millimeters a year, which is as

Eroded landscapes were created by processes whose speeds take away mostly millimeters of surfaces per year.

much as ten orders of magnitude slower than slow walking—a difference that is far outside of our comparative capabilities.[4]

Life appeared early on the Earth, but its subsequent evolution was marked by exceedingly prolonged periods of stagnation and no, or marginal, differentiation. As a result, most of the changes that resulted in today's world of more than 1 million described (and many more millions of surmised) species have taken place during a time span that has extended across only about 10 percent of the Earth's history. Moreover, the most recent acceleration in the evolution of complex life—the emergence of hominins, the group comprising our ancestors (after their lineage diverted from that of chimpanzees) and our sapient species, which led eventually to a global civilization—began about 6 million years ago, or put another way, the most recent 0.13 percent of time that has elapsed from the formation of the planet: geologically speaking, human evolution has unfolded at a very rapid speed.

Of billions and millions: Earth's eons, eras, periods, and epochs

Before getting to the speeds with which the planet changed from a boiling ocean of magma and an atmosphere made up of water and carbon dioxide, to one where water dominates its surface, millions of species inhabit every niche of the biosphere, and one species has achieved an unprecedented (constructive as well as destructive) primacy, I must first describe how the planet and the star around which it orbits were formed. Since the mid-20th century, scientists have been able to put together a coherent model (based on measuring the decay of radioactive isotopes and dating the oldest available minerals) that explains the formation of the solar system. This model indicates that the process was, not only in astronomical terms but even when compared to subsequent evolutions of the Earth's surface and life, rather rapid.[5]

At the beginning there was a large and dense interstellar molecular cloud. Then, starting about 4.6 billion years ago, its contraction (gravitational collapse) and accretion formed a central star and a rotating nebula, a disk made up of a mixture of gases and dust. Through gravity-driven collisions, this dust was aggregated to form progressively larger bodies until eventually it produced planetesimals, objects of about 1 kilometer across; in turn, their collisions produced about 100 much larger (at least Moon-size) planetary "embryos" whose eventual encounters over tens of millions of years formed the planets as we now know them, and one of those large embryos collided with the early Earth and produced its Moon.

In 1956 Clair Patterson, an American geochemist, concluded that "the age for the earth is the same as for meteorites. This is the time since the earth attained its present mass"—and put it at 4.55 billion years.[6] Later studies confirmed and refined this dating. Measuring the decay of uranium and lead isotopes puts the age of the solar system at 4.567 billion years before present.[7] Subsequent transformations followed in comparatively rapid order. The proto-Earth began to form perhaps as early as 3 million years later, and its accretion could have been largely accomplished in just 30 million years; but this timing is uncertain and the Moon-forming collision could have taken place about 100 million years after the formation of the solar system.[8]

Additional mass was gained by further collisions, and the proto-planet might have attained as much as 90 percent of its current mass by 4.45 billion years ago; it became differentiated by the formation of a core made of liquid heavy metals and a lighter mantle and crust.[9] This time span, enormous as it is in absolute terms, appears rather brief in comparison with not only universal history, but also many of the Earth's later extended periods of minimal change.

Now we enter the realm of geological dating, whose primary division of the planet's history is into four eons of unequal duration: Hadean, from 4.54 to 4 billion years ago; Archean (the second longest), from 4 to 2.5 billion years ago; Proterozoic (the longest one), from 2.5 to 0.541 billion years ago; and, until now, the Phanerozoic eon, so far as long-lasting as the Hadean.[10]

No rocks have survived from the Hadean eon (the name referring to the ancient Greek god of the underworld), and so all we can do is make a plausible reconstruction of the primordial conditions of the early Earth—a planet born dry with no water and no atmosphere, and with a magma ocean that lasted up to 100 million years and whose solidification produced the first continents.[11] But it is likely that the oceans formed within 150–250 million years after the Earth's initial accretion, perhaps most likely by water brought by collisions with meteorites that could have delivered a total volume equal to up to three of today's oceans.[12]

The Archean eon lasted 1.5 billion years and the first life forms appeared sometime between 3.5 and 3.8 billion years ago, but possibly

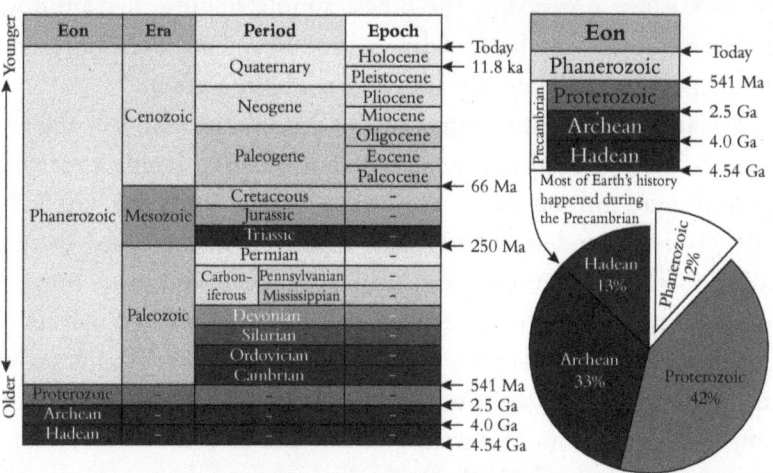

Eons, eras, periods, and epochs: speed on geological scales.

before that time.[13] Seafloor hydrothermal vents seem to be the most likely candidates for the earliest appearance of chemoautotrophic microbes (oxidizing inorganic compounds), but we will never know because there are no preserved materials containing any recognizable living structures older than 3.5 billion years.[14]

Some of the best-preserved early Archean rocks that have been least affected by subsequent high temperatures and pressures are in Western Australia's Pilbara region, where they range from volcanic and sedimentary deposits to those formed by hydrothermal vents.[15] Stromatolites—layered sedimentary rocks that trapped and preserved often-extensive microbial mats—go back to 3.43 billion years ago. William Schopf, an American paleobiologist, compiled the best evidence of their formation from nearly 50 Archean deposits containing biogenic stromatolites, and it strongly supports the conclusion that simple life forms have been around for more than 3.5 billion years.[16]

I should note that, in 2017, a group of scientists claimed that their dating of tiny tubes and filaments preserved in rocks found in northern Quebec put the age of such "putative fossilized microorganisms" at no less than 3.77 billion and possibly as much as 4.28 billion years, a claim that has not been met with universal acceptance.[17] What we know with much greater certainty is that, regardless of the actual age of the earliest organisms, the oldest simple single-celled life forms underwent only very limited diversification and no extension beyond the marine environment for more than a billion years.

The most significant change of the Proterozoic eon was the rapid oxygenation of the atmosphere, which took place during a very brief span of 1–10 million years, beginning 2.33 billion years ago.[18] This (still far from understood) Great Oxidation Event raised oxygen's atmospheric concentration to about 21 percent, and it has remained remarkably stable ever since: without it there would be neither any higher forms of life nor the biogeochemical cycles (those of carbon, nitrogen, and sulfur are the most important ones) that enabled the long-term evolution of plants, animals, and humans. Thus the stage was set for the arrival of photosynthesizing plants and oxygen-breathing animals. Filamentous bacteria dated to 2.72 billion years ago could possibly be one of the first multicellular organisms ever detected, but the first conclusive multicellular cyanobacterial fossils are in rocks that are about 2.15 billion years old.[19]

The next "sudden" surprise came only at the very end of the Proterozoic eon, when new soft-bodied and mobile multicellular life forms that needed oxygen for their metabolism left their marks in rocks that are between 575 and 541 million years old. They became known as Ediacaran fauna, after the eponymous hills (north of Adelaide) where their fossilized imprints in slabs of quartzite were discovered in 1946 by Reginald Sprigg, an Australian geologist.[20] Their shapes ranged from disk-like to worm-like, and some appear to have had no later analogs. This first notable diversification of organisms was almost immediately followed by a similarly rapid but far more abundant appearance of new life forms at the very beginning of the Phanerozoic eon, in which we still live.

The current eon's onset is dated with the precision of "just" 1 million years, a boundary chosen because the fossil record shows that 541 million years ago complex varieties of life began to appear and evolve at (in geological terms) an unprecedented speed. Geologists use the phrase "Cambrian explosion" ("radiation" is a more restrained term), as so many different bottom-dwelling marine animals (both with and without hard parts) and plants (above all, red and green algae) appeared within just 40 million years, after hundreds of millions of years of little or no notable change.[21] And with new body layouts came new capabilities—as these animals evolved to burrow into sediments, to stalk, and to hunt.

Half a billion years ago, the first organisms began to explore land, with the first plants following within 50 million years. The Paleozoic era lasted until 251 million years ago, and its penultimate period, the Carboniferous, was the time of abundant plant cover with massive trees growing in densely vegetated swamps: this exceptionally intensive production of new biomass (mostly cellulose, hemicellulose, and lignin) became buried under sediments and gave rise to most of the world's extensive coal deposits. But the first flowering plants arrived only some 130 million years ago.

No other period in the Earth's geological history has received so much popular attention as the Jurassic and Cretaceous periods of the Mesozoic era (199.6 to 65.5 million years ago). During that time the long evolution of sauropsids (egg-laying vertebrates) made many species of dinosaur the dominant animals in various ecosystems—and their variety (including the largest-ever terrestrial vertebrates), multiple unique body plans (from extremely elongated necks to animals armed with bony plates), and range of ways of life (adapted to activities from swimming to flying and from browsing tall trees to the pursuit of fairly large prey) have become a lasting source of public interest—and commercial exploitation—ever since the parts of giant skeletons were first unearthed during the 19th century.[22] This domination ended abruptly during a major species extinction that separates the Cretaceous and Tertiary periods (widely referred to as the K-T boundary): the Earth's encounter with an asteroid 66 million years ago (with Yucatán's Chicxulub crater identified as the area of impact) is now widely accepted as its cause.

Although the earliest primates had already appeared before the K-T extinction event (they were, most likely, the descendants of small, nocturnal insect-eating tree shrews and flying lemurs that are still around), the lineage (sequence of species) that led to modern humans separated from the one leading to orangutans some 13 million years ago, from the one leading to gorillas 12–8.5 million years ago, and from the lineage that ended in chimpanzees and bonobos just 7–5.5 million years ago.[23] The Pliocene began a few hundred thousand years later, and it ended about 2.6 million years ago. During that time hominins evolved away from Australopithecines (first at about 4 million years ago; small, walking upright but with chimpanzee-size brains) to still ape-like (but with a larger brain) *Homo habilis* using stone tools.

The Pleistocene, the first epoch of the Quaternary period, extended until the end of the last Ice Age, 11,700 years ago, and was followed by the still-ongoing Holocene epoch. By the Holocene's beginning, *Homo sapiens* hunters and gatherers were approaching the time of the most fundamental shift in their lives: their gradual transition to a sedentary existence, supported by domesticated crops and animals. *Homo erectus*, with a brain size at least twice that

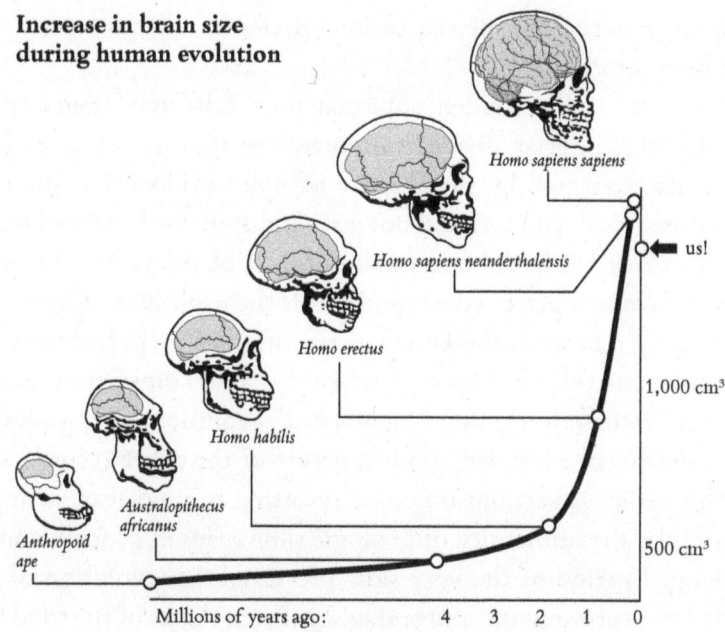

Increase in brain size during human evolution

Homo sapiens sapiens

Homo sapiens neanderthalensis ◄— us!

Homo erectus

1,000 cm³

Homo habilis

Australopithecus africanus

500 cm³

Anthropoid ape

Millions of years ago: 4   3   2   1   0

Exponential sapience: growing brain size in hominins.

of chimpanzees, appeared about 1.9 million years ago, and the oldest known *Homo sapiens* fossils, from northern Africa, are about 315,000 years old, with migrations across Asia and Europe having followed shortly afterwards.[24]

Looking at this long record of the planet and evolution of life, we must note the repeated periods of (comparatively) rapid fundamental advances followed by lengthy periods of slow progress, stagnation, or even retrogression (including five major extinction periods). The formation of the planet took most likely less than 100 million years, the Great Oxidation Event no more than 10 million years, the Cambrian explosion only about 20 million years, and the separation of hominins from other great apes and the evolution of *Homo sapiens* less than 15 million years. If these spans of repeated rapid advances are correct, then they took, respectively, just tiny fractions of the entire length of evolutionary history. If we start counting at 4.5 billion years ago, then the Earth's formation lasted only about 2 percent, the Great Oxidation Event only about 0.2 percent, the Cambrian explosion less than 0.5 percent, and the rise of our species less than 0.3 percent of that long time span.

But what if the sudden appearance of Ediacaran fauna or the rapidity of Cambrian diversification are just illusory—just misleading artifacts created by the absence of older evidence, as the early organisms, small and soft, did not leave indisputable fossilized traces? On the other hand, the onset and duration of the Great Oxidation Period are much better constrained, as is the evolution of hominins, confirming that some fundamental evolutionary steps had relatively abrupt onsets followed by comparatively rapid completion. And in contrast to these intriguing "high-speed" evolutionary episodes, we can point to the extended, gradual nature of the entire record.

This reality is best illustrated by resorting to time-span analogs.[25] Inevitably, the immensity of geologic time spans in general, and the prolonged period of the very slow pre-Cambrian evolution of life, has led to expressing successive stages either as shares of the total time elapsed since the formation of the Earth or as fractions of an hour or a day. Any one of these analogies is especially helpful in emphasizing the duration of the first three eons—4 billion years out of the total of 4.5 billion, or just over 88 percent of the Earth's history. And with the last 541 million years divided into 3 eras and 12 periods, it is not surprising that none of these periods has taken more than 1.7 percent of the elapsed time and that most of them lasted for less than 1 percent of the Earth's evolution (the Cretaceous with 1.7 percent was the longest one, the Silurian with 0.6 percent the shortest). Obviously the most recent Cenozoic epochs take up even smaller fractions of the total—0.5 percent for the Eocene, 0.1 percent for the Pliocene and Pleistocene, and 0.0002 percent for the Holocene.

I find the 24-hour analogy to be the best tool. Imagine an observer of the Earth's life who begins his day-long vigil (initially encased in a controlled cocoon containing oxygen and maintaining a comfortable temperature) near what will eventually become a seashore exactly at midnight; this is analogous to 4.5 billion years ago, when the planet went through the final stages of its near-complete accretion. The first unicellular organisms, the precursors of today's bacteria and archaea, do not appear until 4 in the morning. Then comes a prolonged period

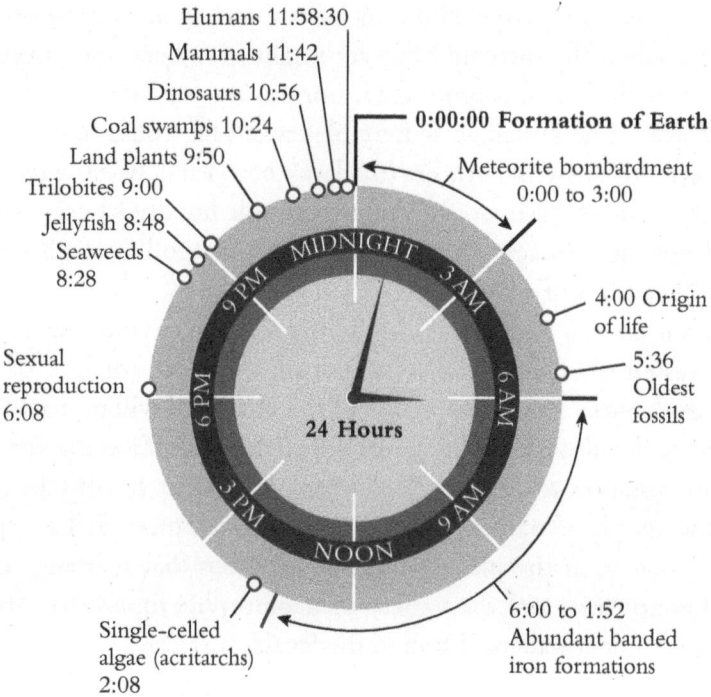

Humans 11:58:30
Mammals 11:42
Dinosaurs 10:56
Coal swamps 10:24
Land plants 9:50
Trilobites 9:00
Jellyfish 8:48
Seaweeds 8:28

0:00:00 **Formation of Earth**

Meteorite bombardment
0:00 to 3:00

MIDNIGHT
9 PM
3 AM
6 PM
6 AM

24 Hours

3 PM
9 AM
NOON

Sexual
reproduction
6:08

4:00 Origin
of life

5:36
Oldest
fossils

Single-celled
algae (acritarchs)
2:08

6:00 to 1:52
Abundant banded
iron formations

A long "day": most of the events that created our environment took place during the last 10 minutes of the planet's history.

of inaction until the Proterozoic eon starts about an hour before noon (10:57:23), and it lasts for more than ten hours, until mid-evening (21:10:20). Oxygen concentrations rise significantly after 12:00 and our observer can now leave his protective bubble, and wading in shallow water might bring encounters with some small soft creatures or the sight of clump-forming microbes. But the Cambrian explosion of diverse marine species would take place only at 21:00, when trilobites buried in the sand would be, even with moonlight, first felt underfoot before being lifted for inspection.

The first land plants would appear in total darkness at 21:49:46; bodies of the first dinosaurs could be silhouetted against the moonlight almost an hour later, at 22:41:30; the earliest small mammals would be seen furtively running around by 23:42:30; the first hominins could inspect the shore only about one and a half minutes to midnight (23:58:30). The first known *Homo sapiens* (who left a skeleton in Morocco) would walk by at about six seconds before midnight, when the intrepid observer might have been long overcome by sleep after such a protracted, boring, no- or low-activity vigil. And even if he remained acutely observant, he would not be able to discern the entire historic era (the last 5,000 years), because its analog on the 24-hour basis would last just a couple hundredths of a second, and our modern age (after 1500 CE) just a few milliseconds (an eyeblink is a third of a second).

Such are the realities of the Earth's and life's evolution seen in a bare-essentials narrative that focuses on the speed (or lack of it) of these advances. In the last section of this chapter, I will return to cover in some detail speeds in life's evolution (from speciation to extinction, from gestation to demise), but before doing that I will take a look at the speeds of natural processes whose outcomes set the stage for life—that is, at the planet's tectonic activities that rearrange oceans and continents, and the variety of geomorphic forces that give the final yet impermanent shapes to the Earth's surfaces.

Sideways, up, down, and away: tectonic plates and geomorphic processes

A coherent theory of the planet's tectonic plates began with the idea of continental drift, first formulated by Alfred Wegener, a German climatologist, early in the 20th century and discounted by most geologists until the early 1960s, when several pieces of incontrovertible evidence fell into place.[26] The first key contribution came in 1962 from Harry Hess, a professor of geology at Princeton (and a former Navy officer), who postulated that ocean floors grow as molten basalt comes up from the Earth's mantle (a nearly 3,000-kilometer-thick layer of solid but hot rock that accounts for most of the Earth's mass) along mid-ocean ridges (that are, significantly, about 1.5 kilometers higher than the surrounding flat abyssal plain), and spreads away in both directions while cooling, subsiding, and eventually, at the deepest ocean trenches (as deep as 11 kilometers, mostly close to continental margins), returning to the mantle to be eventually recycled as new ocean floors.[27]

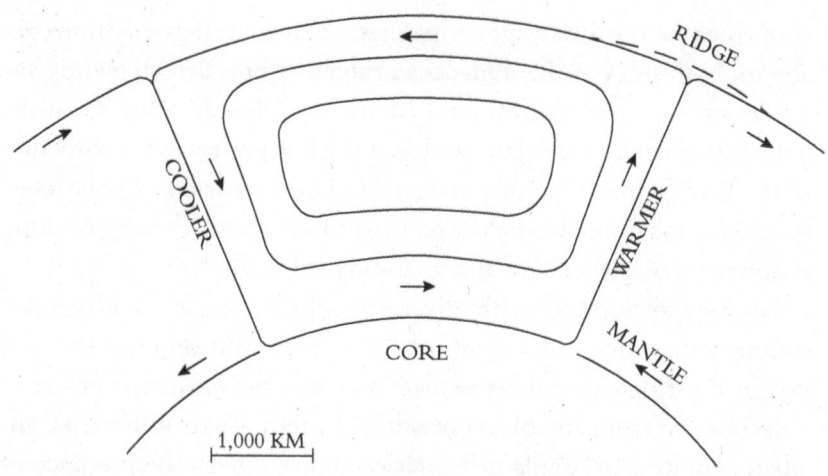

Massive plates adrift: a schematic illustration of plate tectonic movements from
Hess's pioneering publication.

The asthenosphere—the mantle's uppermost zone, made up of
hot and semi-solid material—supports tectonic plates: large slabs of
solid rock that, despite the often-used verb, do not float on it. The
youngest oceanic plates are only about 15 kilometers thick but, being
mostly basalt, they are heavier than the oldest continental plates,
which are up to 200 kilometers thick but composed of lighter minerals. Both kinds of plates have lower density than the asthenosphere,
but their enormous masses make them deeply anchored in the mantle
(and most tectonic plates have both oceanic and continental parts).

Yet another advance took place in 1962 when Robert Coats (at
the US Geological Survey) published the first sketch of a zone where
the oceanic plate is recycled into the mantle.[28] The descent of the
lithosphere (the Earth's outer rocky layer) into the mantle became
known as the subduction process, and subsequent research identified
some 55,000 kilometers of convergent plate margins where descending plates generate most of the power for plate motion and are also
responsible for most of the world's volcanic eruptions and earthquakes.[29] In the long run, the speed of subduction (crust destruction,
or plate recycling) must be the same as the speed of crust creation.

This coherent explanation of the Earth's grand tectonics was soon confirmed by the discovery of magnetic stripes arranged symmetric-ally on both sides of the mid-ocean ridges by two British geologists, Frederick Vine and Drummond Matthews.[30] Basalt rising from the mantle is strongly magnetic, and as it cools it preserves the direction of the Earth's prevailing magnetic field (which switches its polarity—as many as 100 times during the past 20 million years), thereby creating symmetries on either side of a spreading ridge.

Seafloor spreading, with diverging plates, was thus a demon-strable reality, not just an improbable theory. But one fact that did not fit the ridge/spreading explanation was the existence of active volcanoes far from any plate boundary. In 1963, Tuzo Wilson, a Can-adian geophysicist, explained their existence due to the presence of "hotspots"—regions where magma, hotter than its surroundings,

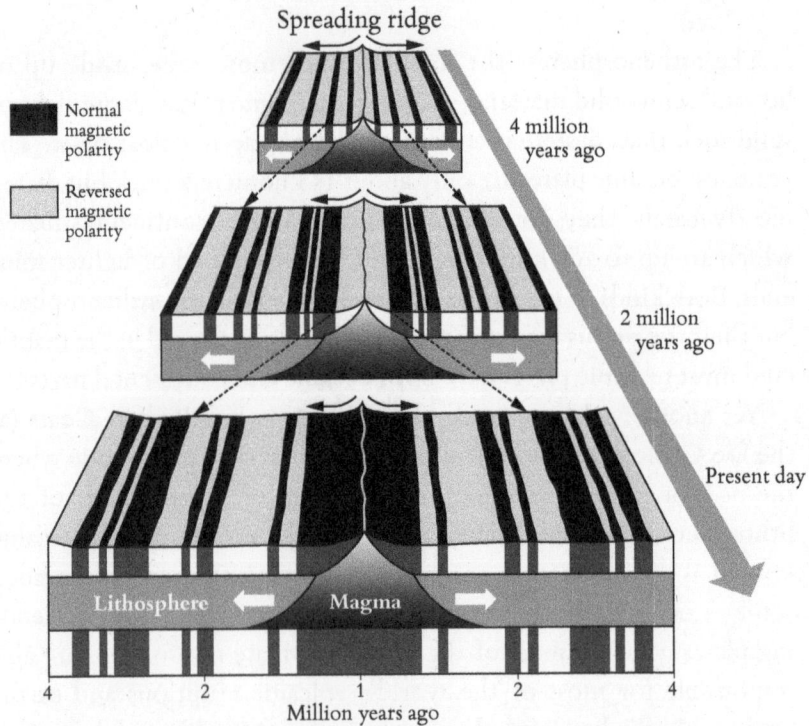

Symmetric seafloor spreading: how Vine and Matthews provided physical proof of the speed of plate tectonics.

rises from the mantle, pierces the crust, and causes episodic volcanic activity. The Hawaiian island chain is this phenomenon's most impressive illustration: as the Pacific plate moves across a hotspot (located undersea, just off the Big Island), it has created a sequence of volcanoes—ranging from long-extinct and submerged peaks northeast of Kauai to Mauna Loa (the world's largest volcano that keeps on, repeatedly, spewing lava) and to Kamaʻehuakanaloa Seamount, a new volcano growing underwater some 35 kilometers southeast of the Big Island.[31]

And in 1965 Wilson identified transform faults as the third kind of tectonic plate boundary.[32] These faults allow the tectonic plates to slide past each other horizontally—that is, without either forming or destroying new crust—and they can be completely submerged, on land, or extend from ocean to land. California's San Andreas Fault, part of the western boundary between the massive North American and Pacific plates, is perhaps the most famous (and most investigated) instance of this arrangement: it runs inland from southernmost California up to San Francisco, and from there it hugs the coast before veering westward at Cape Mendocino, south of the Oregon border.[33]

We now know that the Earth's crust is made up of seven major, but variously sized, tectonic plates (African, Antarctic, Eurasian, Indo-Australian, North American, Pacific, and South American) that are driven by seafloor spreading and recycled into the Earth's mantle by slab subduction; as well as 15 minor plates (including the Arabian, Caribbean, and Nazca plates) and a few scores of microplates.[34] The mantle's asthenosphere is a passive underlay for the grand motions, as the tectonic plates move only when pushed (by mid-ocean spreading) or pulled (by plate subduction); they are not dragged. Subduction is the prime mover of plate motion; ridge push is a secondary force. The Pacific Ocean has a subduction ring that extends in the western hemisphere from South America to Alaska and in the eastern hemisphere from Kamchatka to New Zealand. During subduction the upper plate gets thicker and mountain ranges are formed: the Andes or the chain of spectacular Kamchatka volcanoes are prime examples of this process.

The most notable consequence of the collision of the Indian

and Eurasian plates was the creation of the world's tallest mountain chains—the Himalayas, Hindu Kush, and Pamir, and the more northerly ranges of Kunlun, Tian Shan, and Altai—as well as the world's highest, and most extensive, elevated plateau: if Tibet (with an average altitude of 4,500 meters above sea level) were independent, it would be the world's 10th-largest country.[35] Continental plates move slower because of their larger mass and deep "keels" anchored in the mantle, as do all plates (continental or oceanic) that are hemmed by encircling ridges whose counter-movement results in a kind of stalemate: that is why the African and Antarctic plates are relatively sluggish.

Now, with the grand tectonic features explained and with the grand plate cycling outlined, we can turn to its speeds. As with nearly all other dynamic phenomena, there is a range but a necessarily restricted one. Plates that are some 200 kilometers thick below the continents and about 100 kilometers thick below the oceans, and that occupy areas of tens of million square kilometers (major ones range from 103 million square kilometers for the Pacific plate to nearly 44 million square kilometers for the South American plate), have total volumes on the order of 10 billion cubic kilometers. With a typical crustal density of 2.7 t/m^3, their mass is on the order of a few tens of exatons (10^{18} tons). Given this reality, the fastest inferred—and now also directly recorded—speeds are quite impressive.

Many models of surface kinematics have tried to reconstruct Phanerozoic plate motions, and a reconstruction covering the last 200 million years indicates a median speed of about 40 mm/year across that entire time span.[36] The Phanerozoic maximum was close to 20 cm/year for the Indian ocean plate moving northward between 80 and 50 million years ago. A detailed reconstruction of the convergence history between the Indian and Eurasian plates shows that 72–68 million years ago, before the Indian plate began its accelerated northward motion, the convergence speed was just 8–9 cm/year; after a rapid acceleration the speed peaked at about 180 cm/year; it dropped rapidly 66–63 million years ago, to just above 12 cm/year by 60 million years ago; and after a brief partial rebound by 50 million years ago, it had declined to around 50 mm/year when the continental

masses collided about 45 million years ago.[37] Near Taiwan the convergence speed between the Philippine Sea plate and the massive
Asian plate is about 70 mm/year—and perhaps even faster, up to 82
mm/year (that is, 82 kilometers per million years)—with part of that
convergence taken up by the shortening of the Philippine Sea plate
and the Coastal Range of eastern Taiwan.[38]

Two generalizations are valid: all of today's fast-moving plates,
including the massive Pacific and Indian plates and the small Cocos
and Nazca plates, have no, or only small, continental fractions and
are attached to subducting slabs; all plates made up of large portions
of continental lithosphere have slower speeds, and those without
any subducting margins are the slowest, or immobile. To put this
into a spatial-temporal perspective, moving 10 cm/year is equal to
traversing 100 kilometers—or 1 degree of longitude at the equator—
per million years. A general conclusion is that the oceanic plates are
moving 2–3 times faster than continental plates that are anchored in
the more viscous mantle.

Geologically short-lived accelerations (up to about 10 million years)
took place in the African, North American, and Indian plates—the
last time about 65 million years ago. Recently, the highest equatorial

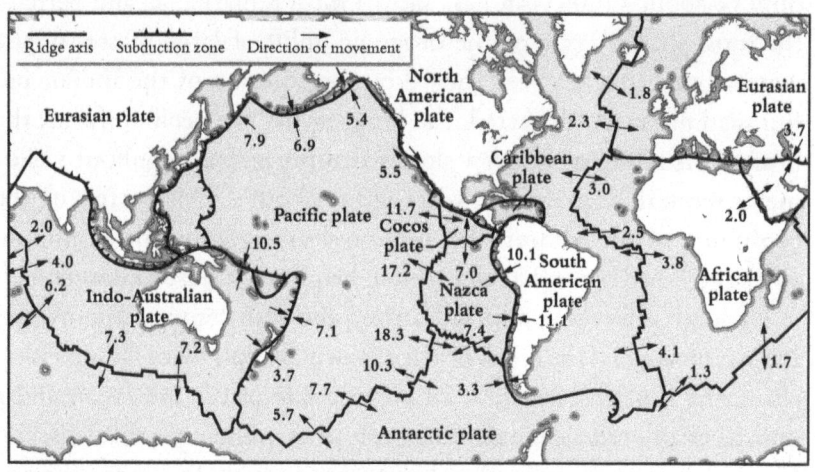

When 15 cm/year is a high speed: the typical annual motions of oceanic plates.
The San Andreas Fault moves 5–7 cm/year.

plate velocity has been 22 cm/year, with the ocean spreading rates ranging from 18 cm/year for the Pacific plate to about 1 cm/year for the large Antarctic plate.[39]

And plate speeds can now be directly monitored, not merely inferred, because they can be highly accurately measured thanks to a network of satellites. Satellite laser ranging that uses the Global Positioning System (consisting of a fleet of satellites, in full operation since 1993) and ground stations on different continents makes it possible to locate precisely points on the Earth's surface, and then, as the plates move, measure the change in the ground stations over a year or a period of years. These accurate measurements show the zone of maximum seafloor motion between 110° and 130° East—roughly between the longitude of Java and the Ryukyu Islands, where the Pacific plate plunges under the massive Eurasian plate at an average speed of 70–72 mm/year (and a maximum of up to 80 mm/year).[40]

But colliding plates can also bring about the very opposite of subduction: the formation of mountains (orogeny) whenever the speed of mountain uplift surpasses the speed of wearing down the rocks by their continuous gravitational collapse and erosion. Uplifting caused by plate convergence in East Asia could be 7.5–17.5 mm/year in Taiwan's easternmost Central Range, but longer-term rates might only be about 5 mm/year.[41] By far the most remarkable, and perhaps the most studied, case of the orogenic uplift of large masses of the continental crust has been caused by the collision of the Indian and Eurasian plates. As expected, the progress has not been uniform: the proto-Himalayan range grew slowly from an altitude of about 1 kilometer more than 40 million years ago to about 2.3 kilometers by the beginning of the Miocene (23 million years ago), but by 11 million years ago it had more than doubled in height to at least 5 kilometers.[42]

Geodetic surveys conducted during the 20th century in northern India indicate that the block north of the main boundary fault is rising above the southern block at an average rate of 0.8 mm/year, and a multitude of evidence (including sediments, pollen, plant fossils, and traces of ancient snowlines) makes it safe to conclude that the average speed of uplift since the late Pliocene (about 2.6 million years ago) has been about 1 mm/year (1 cm/decade or 1 km/million years). Zones that have seen higher uplift include Nepal's Greater Himalayas, with localized speeds of about 4 mm/year, and Lesser Himalayas, with a maximum of about 2 mm/year.[43] Radar measurements indicate a maximum of 7 mm/year at the front of the Annapurna mountain range in Nepal, while data from southern Tibet show uplifting exceeding 1 km/million years between 17 and 11 million years ago, followed by a slower but continuing rise.[44]

These reviews of uplift speeds could be extended to other plate collisions, most dramatically to that of the oceanic Nazca plate and South American plate that has been responsible for the Andean uplift, and to the collision of the African and Eurasian plates that created the Alps. The Andean uplift remains very active: for example, shell materials from elevated beaches in the Chiloé region indicate a speed on the order of 1 cm/year during the past 3,000 years[45]— while the Western Alps present an example of spent orogeny. GPS measurement shows no horizontal plate motion in the latter region (the African and Eurasian plates are motionless along that contact zone), but they do indicate a regional pattern of uplift amounting to about 2.5 mm/year (25 cm/century) in the northwestern Alps: given the absence of any active plate convergence, the observed mountain uplift must be due to deep-seated changes.[46]

In any case, long-term speeds of uplift cannot be calculated by a naive division: for example, the tallest Himalayan peaks are now close to 9 kilometers above sea level, and if assuming the Indian plate began lifting the Eurasian plate about 80 million years ago, a simple division would imply an annual uplift speed of 0.1 mm/year. But that would ignore denudation, the other ever-present category of geomorphic forces, which acts in the opposite direction to uplift. The Earth's surfaces are formed by tectonic processes taking place inside the Earth, driven mostly by the primordial heat that is slowly transferred from the core to the mantle and the crust, and is accompanied by volcanism, earthquakes, and landslides. Subsequently, the planet's surfaces are re-formed by denudation (collapse, weathering, and erosion), as well as the downslope and downstream transportation of displaced materials and their eventual deposition in sedimentary formations.

These forces of denudation—dominated by gravitational collapse, the kinetic energies of water and wind, and the abrading flows of glaciers, but also including the freezing and melting of water in soil and rock cracks, the actions of plant roots and subterranean organisms, and chemical reactions affecting soils and minerals—proceed at speeds that can match the speed of uplifting or can be, during some (even protracted) periods, considerably slower or faster. Obviously, these external geomorphic forces have been particularly intense in high-mountain environments that are made up of complex rock formations, glaciers, steep slopes, and plunging streams, receive heavy precipitation, and are subject to considerable temperature extremes.

Moreover, and no matter how counterintuitive this seems, in these high-mountain environments erosion is also a process that contributes to mountain uplift.[47] That reality is determined by the basic physical property of tectonic plates. Because the density of continental crust is 15–20 percent lower than the density of the underlying mantle, mountain ranges—much like icebergs (with ice being 10 percent lighter than water), which are partially above water but mostly (90 percent) below the waterline—are partially above the mantle but mostly (at least 80 percent, or typically more than 10 kilometers) rooted in it. As erosion strips away large masses of rock and rivers move the eroded material away, the denuding ranges become lighter and hence tend to rise as isostatic forces (maintaining the equilibrium between the crustal and mantle masses) lift the crust to replace most of the mass that has been stripped and transported away. As a result, some 20 kilometers of rock would have to be stripped away to completely flatten a mountain range that was 5 kilometers tall to begin with.

Isostatic forces are also very much in evidence in many flat landscapes that were covered by massive ice sheets whose latest maximum was reached about 20,000 years ago: their melting initiated the slow process of post-glacial rise that is still going on in the tectonically stable regions of Scandinavia and North America. For example, the fastest rise in Canada is taking place along the southern shore of Hudson Bay in northern Ontario, where the land is rising by 10–13 mm/year (1–1.3 m/century).[48] Moreover, a recent assessment of geodetically measured uplift in the Alps showed that the retreat of much smaller Last Glacial Maximum icecaps can still dominate the present uplift.

Mountain building is thus more than a simple process of tectonic uplift; it is a complex combination of forces that elevate the rocks, denude them, and compensate for the denudation, with erosion and isostasy being positive forces in the long run. Moreover, the outcome of this process has profound effects on climate, and these changes also contribute to the speed of orogenic and denudation processes. Perhaps most notably, Himalayan uplift has elevated the large Tibetan plateau while also blocking the monsoonal flows from reaching it, a combination that has slowed down its erosion.[49] But erosion's effects can be seen in every environment, with human activities (particularly poor farming practices) often accelerating erosion losses.

How fast does water evaporate and precipitate?

Before quantifying these erosion speeds, this is perhaps the most apposite place to look at the speeds of water and air, the two dominant geomorphic agents of erosion and denudation and, of course, two media inhabited by organisms whose speeds I will survey in the next chapter. I will start with rain, but that means starting with evaporation and condensation—getting enough water into the atmosphere, and converting it from vapor to liquid. Temperature is the decisive factor for the speed of evaporation, as it imparts sufficient kinetic energy to individual water molecules to escape from the liquid state. Above boiling point (100°C), all water molecules have sufficient speed to escape from the liquid phase—while at any lower temperatures this is only the case for those molecules attaining that critical speed: some can do so even at the point of water becoming frozen (0°C), but evaporation at low temperatures proceeds very slowly.

In addition, evaporation speed increases with a larger surface area (evaporating seawater in shallow coastal ponds exploits that principle for harvesting salt) and with wind speed, and it decreases with higher atmospheric pressure (water will evaporate faster in Tibet, 3.5–4.5

kilometers above sea level, than in New York or Rome, but how much faster will also depend on actual temperature, humidity, and wind speed) and with air humidity (evaporation is rapid in deserts, and slow in tropical rainforests). If you are curious how much water you lose in summer to evaporation from a backyard pool, you can get the answer by plugging basic variables into a variety of available formulas.[50] For example, during the summer in Texas a swimming pool with an area of 75 square meters (15 × 5 meters) will lose about 700 liters of water every day to evaporation—an amount about twice as much as the average US household water use per capita!

Matters become far more serious for operators of large hydro stations in hot climates. None of those is more affected than the reservoir created by the Aswan High Dam in southern Egypt, located in one of the world's hottest desert regions. Studies during the past 50 years have shown a minimum January evaporation of about 4 mm/day and summer maxima above 10 mm/day, adding up to a total annual water loss of nearly 15 billion cubic meters, or about 12 percent of the lake's total water content.[51]

Condensation is the opposite process, as cooling (often with the help of small airborne particles) returns vapor to liquid. A giant summer thundercloud (cumulonimbus) is the most accelerated example of condensation, as updrafts with speeds of more than 50 km/h move the air to heights 10–15 kilometers above ground, preliminary to torrential rain. And with what speed do those thundercloud raindrops fall? As with any falling objects released from high above, the speed of raindrops increases until it reaches its terminal velocity, which is size-dependent. Large raindrops (about 6 millimeters in diameter) will reach a maximum speed of about 10 m/s; this means that a raindrop formed inside a large summer cumulus cloud some 5 kilometers above ground would take at least 8 minutes and 20 seconds to hit the ground—but if it gets caught in an updraft it may not fall for tens of minutes or not at all (it may evaporate).[52] Small raindrops (3 millimeters in diameter) cannot fall faster than 7 m/s, and drizzle (just 1 millimeter across) falls at just 2 m/s.

Kinetic energy is equal to one-half of the product of the mass and the speed squared ($E_k = 0.5mv^2$), and this means that, everything else being equal, doubling the speed of raindrops will quadruple their energy and hence their erosive capability as they impact an unprotected surface—above all, agricultural soils. As a result, rain-driven soil erosion of farmland is particularly intense in regions receiving heavy seasonal downpours, especially in monsoonal Asia where even faster erosion comes from precipitation's runoff (and worst of all, from flash flooding). Precipitation eventually ends up in streams, and water erosion caused by stream flows (both in their normal channels and when flooding) is an incessant process—as is, in many regions, wave action along coastlines.

Water flowing down steep mountain inclines moves fast: when cascading over large boulders and free-falling more than 5 meters, its speed will surpass 10 m/s. Once the streams leave the mountains, they slow down considerably as their gradient (the steepness measured in degrees as a river flows downstream) declines.[53] The Amazon has the flattest course of all major rivers, descending only 90 meters over some 3,200 kilometers of its journey to the Atlantic Ocean. The speeds of major rivers are mostly between 1.4 and 2.8 m/s (5–10 km/h)—the Amazon's average is about 2.4 km/h—and they increase in narrows and during floods. Some rivers are so slow when entering the sea that they can experience reverse flows as tidal waters move upstream: in the Amazon, the tide's effects can be seen nearly 1,000 kilometers inland, and the *pororoca* tidal bore that is driven by spring tides can move upstream from the river's mouth at speeds up to 24 km/h, creating a wall of water up to 4 meters tall.[54]

And water also flows in its solid state as glaciers move downslope. The fastest-moving ice streams that are parts of Greenland's or Antarctica's ice sheet have surprisingly high speeds, in the range of 6–35 meters a day.[55]

Still, even the record distance of 12.6 km/year implies a speed of 0.0004 mm/s, or four orders of magnitude slower than typical stream flows. And with global warming there has been much more concern about retreating, rather than advancing, glaciers, particularly in the Alps. Between 2000 and 2014, the overall loss was about 39 square kilometers, and the mean elevation retreat was 60 cm/year (with maxima up to 90 cm/year).[56] This means that the retreating glaciers,

A river of ice: a rapidly flowing Greenland glacier.

besides storing less water, are also becoming less important agents of mountain denudation as they cease to form large lateral and terminal moraines (rock debris deposits) and glacial lakes.

Wind and water: sculpting our world

Moving water is the dominant agent of erosion and denudation, but the natural process of loosening, removing, and transporting solids (whose sizes range from light soil particles to heavier sediments and rock fragments) is also done by gravitational collapse (especially in the world's tallest mountain chains), by glaciers forming mountain valleys and abrading bedrock, and by sufficiently strong winds. Wind deflation removes lighter particles, leaving heavier materials behind; in turn, airborne materials can impact solid surfaces and erode them by abrasion. Winds can sometimes carry eroded materials across considerable distances: Saharan sand and dust blown northward (coloring Alpine snow) and transported westward (deposited on Caribbean islands) are notable, and recurrent, examples of such long reaches.[57]

Wind's destructive energies also rise with the square of its speed, but in most instances they are inferior to those of flowing water: water (at sea level) is more than 800 times heavier than air (1,000 kg/m^3 vs.

1.225 kg/m³). This means that 1 cubic meter of a moderate breeze (28 km/h) has an energy of a mere 37 joules, while the kinetic energy of 1 cubic meter of a moderately fast-flowing river (5 km/h) is 965 joules, and that wind would have to reach a hurricane speed (more than 140 km/h) to have the same kinetic energy. That is why relatively slow-moving floodwaters cause far more damage than even strong (but non-hurricane) gales and storms (wind speed categories 9 and 10, between 75 and 102 km/h); only hurricanes (category 2 has speeds above 154 km/h, category 5 above 252 km/h), and even more so tornados (maximum possible short-term speeds on the order of 500 km/h), are near-instantly destructive.[58]

These events, and other most intensive geomorphic forces, can leave behind stunningly altered terrain in a matter of hours—even minutes—but the usual rates of denudation and accretion (erosion and sedimentation) are very slow, with changes apparent only after hundreds, but in most cases only after thousands or millions, of years. Measurements of the decay of a radioactive isotope of beryllium present in quartz have been used to quantify this slow denudation.[59] Their values in the Indian Himalayas imply erosion rates as high as nearly 8 mm/year. The Trans-Himalayan rivers erode the bedrock 0.4–2.5 mm/year, but their different slope profiles result in different speeds of headward erosion (the moving of a river's origin upstream from the direction of its flow): as fast as 90–100 mm/year for the Trisuli and Arun, and 55–60 mm/year for the Marsyangdi and Kali Gandaki, but only 40–45 mm/year for the Budhi Gandaki and Sun Kosi.[60]

Again, geological timeframes are imperative to appreciate these denudation speeds. The just-cited erosion rate of 8 mm/year in parts of the Indian Himalayas means that up to about 8 meters of rock-wall slope erosion would be possible in those regions across a single millennium, and deposits more than 2 kilometers deep would be removed when extrapolated for the entire Quaternary period (the past 2.6 million years). How much lower are the removal rates outside of the highest mountain chains, and what are some typical denudation or erosion speeds for entire watersheds or for major agricultural regions where annual cropping exposes farmland to accelerated soil erosion?

Various measurements and material-flow budgets show that tectonically stable areas, no matter if they are in wet tropics or in arid subtropics, lose only a few meters of surface per million years, compared to kilometers in the regions of the fastest uplift, and that the long-term uplift, rather than difference in climate, appears to be the main controlling factor of short-term erosion rates. Estimates for the world's major river basins are available for both mechanical and chemical denudation—based, respectively, on measurements of total sediment and of dissolved materials. Mechanical denudation rates are as low as a truly negligible 0.001 millimeters (or 1 micrometer) per year for the St. Lawrence River watershed, and as high as about 0.7 mm/year (7 cm/century) for the basin of the Brahmaputra, while even the highest chemical denudation rates (about 0.03 mm/year for the Chang Jiang basin) are a small fraction of kinetic losses.[61]

Obviously, the greatest public concern is about the erosion of arable land: it degrades affected soils, reduces their organic matter and water-holding capacity, and lowers crop yields. The most common way to report these losses is in tons per hectare per year. To compare these rates with denudation losses measured in millimeters per year we must assume an average topsoil density: if it were as low as 1.1 tons per cubic meter, then losing 1 ton per hectare per year (generally considered as a tolerable annual loss) would be equivalent to a denudation of 0.09 mm/year or nearly 10 centimeters (approximately the width of an adult fist) in 1,000 years. Actual losses are commonly much higher even in well-farmed regions with winter crops and the frequent cultivation of cover legumes: the recent British average is 1.27 t/ha per year compared to 0.72 t/ha per year for grasslands.[62]

Studies in the US Midwest, where some remnants of the original native (uneroded) prairie sit above the surrounding (and by now significantly eroded) farmland, offer a unique opportunity to assess the total amount of soil lost since the beginning of farming in that fertile region one and a half centuries ago. High-resolution topographic surveys have shown a median reduction in soil thickness of between 0.04 and 0.69 meters: that corresponds to erosion rates of 0.2–4.3 mm/year, with a median historical erosion rate of 1.8 mm/year or nearly double the rate considered tolerable by the US Department of Agriculture.[63]

And a high-resolution, global dataset containing more than 35 million observations shows that prevailing national uses, practices, and policies have a major effect on the speed of farmland erosion. While the study puts the global average soil erosion rate at 2.4 t/ha per year, it estimates that the average difference of intercountry comparison was rather high, at 1.4 t/ha per year.[64] The countries whose national averages are far above the mean departure (with differentials close to or above 3 t/ha a year) include, most notably, Brazil, China, and Mexico, while the countries whose differential is far lower than the global mean (that is, those who farm better) include Argentina, Tanzania, and the US.

So far, I have looked at tectonic and geomorphic processes that proceed at absolutely very slow speeds, ranging from the proverbially glacial (as slow as fractions of a millimeter a year for hard rock erosion; with glaciers, as we have seen, usually much faster-moving) to slow and gradual changes (on the order of a few centimeters a year) that become apparent in landscapes only after many years or decades of gradual uplift or denudation. But there are two notable exceptions that proceed at rapid speeds, two categories of sudden, mass-scale energy releases associated with plate subduction that produce major geomorphic changes: earthquakes and volcanic eruptions.

Earthquakes, volcanoes, tsunamis

Earthquakes are the most frequently felt consequences of plate tectonics. Their magnitude is measured on a logarithmic scale (a magnitude 6 quake is 10 times more powerful than a magnitude 5 tremor), and their frequency shows an exponential decline with increasing magnitude: worldwide, there are hundreds of tremors of magnitude 2 and lower every day; major events (greater than magnitude 7) take place more than once a month; exceptional earthquakes (>8) once a year.[65] Similarly, small, locally confined volcanic eruptions are common; exceptional events are rare. And I must add that some (usually less powerful) earthquakes and volcanic eruptions take place far away from subduction zones: the former are usually associated with deep-seated movements of materials within the upper mantle; the latter (as already noted) with hotspots.

Earthquakes are caused by a sudden rupture along a plate boundary. Ruptures begin to propagate from a point several kilometers below ground and, depending on the magnitude of the event, they can range from just a few to more than 1,000 kilometers: in 2004, the 9.2 Sumatra earthquake rupture ran for more than 1,200 kilometers.[66] The speed with which these ruptures propagate depends mainly on the properties of the rocks that are torn apart. The maximum of the magnitude 7.5 Palu earthquake on Indonesia's Sulawesi in 2018 was thought to be about 11,500 km/h (about 3.2 km/s), but studies have indicated a rupture speed of 4.1 km/s from its initiation to its end— that is, about 12 times the speed of sound.[67] Obviously, these ruptures liberate tremendous amounts of energy—even a Richter magnitude 6 earthquake releases more than 60 terajoules, the equivalent of the Hiroshima bomb—and these sudden releases often cause enormous material damage and casualties and generate seismic (elastic energy) waves.

The frequencies of these waves range mostly between 0.01 and 10 hertz but can go as high as 100 hertz.[68] We can hear sounds with frequencies between 20 and 20,000 hertz, which means that only the seismic waves with the highest pitch are audible. After a rupture, seismographs (sensitive recorders of tremors) will first receive the primary body waves, or P-waves, that go through rocks, ocean, or even the atmosphere—while the secondary body waves, or S-waves, propagate only through solids as they shake the ground through a crosswise motion, up and down, perpendicularly to P-waves.[69] Only then come the surface waves (two kinds, named after the scientists who first identified them, Love and Rayleigh), which are slower but much more destructive.

The speed of seismic waves is on the order of a few kilometers per second, and it depends on the material through which the waves propagate (they move faster through rocks than through water), as well as temperature (a higher temperature lowers the speed) and

pressure (velocity increases with depth, i.e. with higher pressure).
Faster P-waves travel 6 km/s near the surface and more than 10 km/s
near the Earth's core: the latter speed is 36,000 km/h, or about 25 percent faster than the orbiting speed of the International Space Station.
S-waves propagate at 3.4 km/s at the surface and about twice as fast
near the core, through which (as it is liquid) they cannot pass, and the
speed of surface waves is similar: 2–6 km/s for Love and 1–5 km/s for
Rayleigh waves.[70] This means that the first P-waves from an earthquake in the Pacific Ocean just east of Japan will hit seismographs in
California (a distance of about 8,000 kilometers) about 11 minutes
later, with the S-waves arriving 10 minutes after that.

Much like exceptionally large earthquakes, unusually powerful
volcanic eruptions—such as those of Vesuvius in 79 CE and Mount
St. Helens in 1980 (a volcanic explosivity index of 5) or Krakatoa
in 1883 and Pinatubo in 1991 (a VEI of 6)—are rare, and small, limited eruptions with restricted local impacts are common.[71] The best
example is provided by Hawaiian volcanoes: that great hotspot now
generates mostly VEI 0 events, with ejection columns of less than
100 meters and with limited and slow-moving lava (especially the
rope-like pahoehoe) that can not only be outrun but also easily out-
walked. The fastest recorded downslope Mauna Loa lava flows are 2.6
m/s; typical speeds are around 0.03 m/s or 100 m/h.[72]

In contrast, Plinian eruptions (Vesuvius-like, so vividly described
by Pliny the Younger) and ultra-Plinian events (Krakatoa, Tambora)
involve high-speed turbulent flows of gas and magma mixtures with
peak gas velocities being supersonic, often approaching a choking
velocity of 600 m/s.[73] Such eruptions can eject massive volumes of
ash and gases, with columns reaching as high as the mid-stratosphere:
the Mount Pinatubo ash cloud extended up to 35 kilometers above
sea level. Ejected ash and gases (above all, sulfur dioxide) have not
only local and regional impacts (burying vegetation and structures,
and acidifying waters) but also global consequences, as they cause
measurable tropospheric cooling by intercepting some of the incoming solar radiation.

Depending on the eruption's magnitude, heavier volcanic ejecta—
hot lava flows, tephra (rock pieces), lahars (mudflow and debris flows

down the volcano sides) and pyroclastic flows (made of hot volcanic gases and ashes, pumice, and stones)—can be restricted to relatively small areas surrounding an erupting volcano or they can affect areas within a radius of 10–20 kilometers. The two fastest-moving ejecta are pyroclastic flows and lahars. The former are analogous to snow avalanches. They originate either from the collapse of an eruption column or from the collapse of lava accumulation on a steep slope. Dense basal laminar pyroclastic flow is superimposed by the turbulent flow of hot fragments, ashes, gases, and steam whose downslope speed is usually more than 100 km/h and temperature is commonly more than 200°C.[74]

These rapid flows can extend up to 20 kilometers from a volcano but extremes further than 100 kilometers are possible, and their effect may be magnified when they melt snow and ice on the glaciers covering volcano slopes. There is no defense against these flows: they are too fast to escape, too asphyxiating and too hot to survive, killing all organisms within their reach. Events in 79 CE (Vesuvius) and 1902 (Mount Pelée in Martinique) are the two most famous instances: the first one entombed Pompeii and Herculaneum (with a total regional death toll up to 16,000); the second one killed nearly 30,000 people, or 15 percent of the island's residents.[75]

As for the highest estimated velocities, during the explosion of New Zealand's Taupō (about 1,800 years ago), the pyroclastic flow had a very high initial velocity of 500 m/s (and also a very large radial distribution), and it was able to climb many vertical obstacles.[76] The pyroclastic flows that accompanied the famous eruption of Vesuvius had an initial velocity of between 50 and 100 m/s (up to 360 km/h) and a flow thickness ranging between 2 and 10 meters.[77] In 1883, the Krakatoa eruption (just west of Java) discharged about 12 cubic kilometers of dense magma, and its pyroclastic flows were discharged into the sea with initial speeds exceeding 300 km/h and then slowing down to just over 100 km/h.[78] These speeds are often high enough to overcome vertical obstacles of hundreds of meters in elevation and can transform surfaces: the 1980 Mount St. Helens blast (at 360 km/h) was an excellent example of these effects, as it completely devastated a fan-shaped area (roughly 37 × 31 kilometers) and knocked down 600 square kilometers of forest.[79]

Lahars (a Javanese term) are wet analogs of pyroclastic flows, composed of a hot or cold mixture of water with volcanic ash and rock fragments. Depending on their density and on the steepness of the slope, they can flow rapidly (speeds exceeding 200 km/h) or move rather sluggishly, like watery concrete: in the first instance there is no possibility of escape; in the second, and if the flow is not too deep (it can be tens of meters thick), it is possible to survive by seeking higher ground. Lahars pose the worst danger when moving down the steep slopes of stratovolcanoes: the November 1985 eruption of Colombia's Nevado del Ruiz produced two lahars that killed more than 23,000 people.[80]

Both earthquakes and volcanic eruptions can generate tsunamis—extremely long water waves that radiate from their center of origin and can have disastrous impacts on shorelines thousands of kilometers away. While wind-generated waves succeed each other in 5–20 seconds and have wavelengths mostly between 10 and 100 meters (and maxima of around 1 kilometer), tsunamis have periods ranging from 10 minutes to 2 hours and their wavelengths can be more than 500 kilometers.[81] But what makes them so destructive is their speed. A tsunami with a wavelength of 200 kilometers traveling across the Pacific (an average depth of 4,000 meters) will travel, with a minimal loss of energy, at a speed of about 700 km/h, and can reach speeds of about 900 km/h in waters 7,000 meters deep—that is the speed of a jetliner cruising at the edge of the stratosphere!

And it will do so unnoticed by the ships passing above it. Only when it enters shallower coastal waters will its speed decline and the height of its wave increase. Depending on its initial power and speed, as well as on undersea features and the configuration of the coastal zone (the presence of narrows, bays, beach slopes), it may break far offshore and be akin to a rapid but modest tidal wave—or it can arrive as a wall of water 10–15 meters tall (even 40 meters if it is generated near an earthquake's epicenter) and proceed to flood (again, depending on the terrain) an area from tens or hundreds of meters to a few kilometers inland, bringing destruction in its path, covering the land with debris, and then carrying some of it away as it recedes.

Two catastrophic events took place in the early 21st century—in Southeast Asia in 2004 and northern Japan in 2011.[82] On December

26, 2004, an earthquake of magnitude 9.1–9.3 (the third-highest ever recorded) struck just off the northern coast of Sumatra, causing a 30-meter tsunami that killed nearly 228,000 people in Indonesia, Thailand, Sri Lanka, and India. The epicenter of the magnitude 9.0 Tōhoku earthquake on March 11, 2011, was just 72 kilometers off Japan's Pacific coast and generated a tsunami whose initial speed was about 800 km/h and generated on impact a wall of water as high as 40 meters, traveling in some parts more than 5 kilometers inland and destroying 450,000 homes. Some 98 percent of more than 18,000 casualties were attributed to the tsunami rather than the tremor itself.

But by far the highest tsunami wave was generated in July 1958 by a 9.1–9.3 earthquake in Alaska's Lituya Bay: it loosened some 30 million cubic meters (90 million tons) of rock high above the inlet's northeastern shore, and as that mass plunged about 900 meters into the waters of the Gilbert Inlet it generated a mega-tsunami as high as 524 meters, destroying all trees growing below that height.[83]

And in 2001, two American geologists published a paper concluding that the densely populated US East Coast could be affected if an eruption of the Cumbre Vieja volcano in the Canary Islands resulted in a massive landslide of volcanic rock and generated a mega-tsunami that could, after crossing the Atlantic, hit Boston, New York, and Miami with waves many meters high.[84] But later investigations showed a low probability of such an event, especially when compared to the danger posed by tsunamis triggered by large underwater earthquakes taking place along plate subduction zones. A sudden upward movement of an overriding plate of the Cascadia subduction zone off the coasts of Oregon, Washington, and British Columbia could generate a tsunami that could hit the shore in just 15–20 minutes, leaving little time for evacuation to escape waves that could be 10–15 (and perhaps up to 30) meters high.[85]

The pulse of life: diversity, speciation, extinction

Diversity is life's most impressive readily observable attribute, with the size range extending from near-microscopic insects (fairy wasps, with bodies no longer than 1 millimeter) to the enormous bodies of blue whales (record size more than 30 meters long), and with mobility ranging from immobile corals (relying on ocean currents to deliver their food) to fast-diving birds of prey. Life's metabolic and reproductive complexities are even more astonishing. The tolerable temperature range is among the most notable examples of these adaptations: the deep-sea worm *Alvinella pompejana* survives near hydrothermal vents at 80°C, while Antarctic fishes live in seawater whose temperature is close to its freezing point (lowered by the presence of salt) of −1.86°C.[86] But the understanding of adaptations needed to cope with such extremes requires reasonable grounding in several scientific disciplines. In contrast, all that is required to be dazzled by life's outward diversity is to look around and see the profusion of forms, sizes, and habits.

Modern taxonomic classification proceeds from species (taking the Eastern grey squirrel as an example: *Sciurus carolinensis*) to genera (*Sciurus*), families (Sciuridae), orders (Rodentia), classes (Mammalia), phyla (Chordata) and kingdoms (Animalia).[87] Some of these groups are astonishingly diverse, while others are limited to just a handful of species, or even just a single species. The family of Asteraceae (asters, daisies, sunflowers) contains nearly 2,000 genera; the genus *Astragalus* (milkvetch) has more than 3,000 species; while monotypic genera (with a single species) include gingko (a large Chinese tree with unmistakably fan-shaped leaves), platypus (egg-laying Australian mammal), beluga whale, and *Homo* (*sapiens* being the only survivor on that evolutionary branch).[88]

We now have a better appreciation of this, still largely unidentified complexity. The classification of life forms began when Carl Linnaeus published his *Systema Naturae* in 1735, and a recent study has shown that subsequent assignments of newly identified species to higher taxonomic groups have followed a consistent pattern from which it is possible to predict the total number of species on land and in the ocean.[89] Described species now number about 1.25 million, and the predicted total is 8.7 million, with 2.2 million living in the ocean. That implies that about 86 percent of terrestrial and 91 percent of marine species are yet to be identified and described.

Whatever the final species count might be, this great diversity of life is the result of past rates of speciation, or the formation of new species—that is, distinct groups of organisms that can reproduce with one another but are unable to do so with similarly sized and almost identically functioning organisms: where I live we have, besides Eastern grey squirrels (some of which can be black), three other species—American red, fox, and northern flying squirrels. Charles Darwin, in *On the Origin of Species by Means of Natural Selection*, concluded (before any genetic evidence became available) that new forms of life can arise over long periods of time by one species splitting into two, or by a population that diverges so much from its extant ancestor (a process that can be accelerated by becoming isolated) that it forms a new species.[90]

A naive estimation of the speciation speed for complex animals and plants would be to start with the Cambrian explosion and divide the number of known (scientifically named) extant species (1.25 million) by the time elapsed (541 million years) to get the average rate of species formation per year. But the resulting speed of 0.002 species formed per million years during the period of more than half a billion years would be a massive underestimation, because all but a tiny share of species that have ever lived are extinct, with most of the losses taking place during one of the five massive extinction events that have swept the biosphere since the Cambrian explosion.[91]

The Ordovician-Silurian extinction (440 million years ago) predated the appearance of terrestrial vertebrates and hence it was limited to marine organisms, with 68–77 percent of all species lost. The late Devonian extinction (365 million years ago) affected many tropical marine species, with 42–69 percent lost. The Permian-Triassic extinction (250 million years ago) was the largest of the five, with a very broad impact as it wiped out 80–86 percent of all marine and 97 percent of all terrestrial species. The Triassic-Jurassic extinction (210 million years ago) opened the way for dinosaurs by eliminating about 70 percent of both marine and terrestrial species, while the Cretaceous-Tertiary (K-T) extinction (65 million years ago) reduced, once again, both marine and terrestrial species by about 70 percent, ended the dominance of non-avian dinosaurs, and marked the beginning of the ascent of mammals.

The end-Cretaceous mass extinction is not only the most famous (thanks to the final demise of the dinosaurs) but is also among the most destructive events of all time. A recent re-examination of mammalian diversity shows that it was more severe than previously acknowledged—but that the recovery of mammalian species was also more rapid than previously concluded. In North America the extinction rate was more than 90 percent—but within 300,000 years local diversity had recovered and regional diversity was twice as high as before the extinction event.[92]

The most recent, and a relatively minor, global mass extinction (not counted among the top five), was at the Eocene-Oligocene boundary 34 million years ago, and it was marked by the onset of Antarctic glaciation. It amounted to about a 15 percent loss of shelly marine invertebrates; and in the event's aftermath, modern whales, seals, sea lions, and walruses diverged from other mammalian carnivores.[93] The most notable prehistoric extinction was the relatively rapid elimination of the Pleistocene (2.58 million–12,000 years ago) megafauna—including, most notably, the woolly mammoth and woolly rhinoceros, as well as other large herbivores, such as the giant deer, steppe bison, auroch, and (in North America) mastodons and giant ground sloths.

In 1861, Richard Owen, an English biologist, came up with the "overkill hypothesis" that attributed the rapid extinction of the North American megafauna to hunters migrating southward from easternmost Siberia, and this explanation was revived during the late 1950s by the American geoscientist Paul Martin.[94] But the megafauna in Africa were hunted long before any humans arrived in North America, without similar destructive effects, and in Europe and Asia hunters and mammoths coexisted for millennia. Undoubtedly, hunting was a factor in the megafauna's extinction in North America, but many studies have identified climate change (leading to the displacement of the nearly treeless ecosystems supporting masses of large herbivores by woodlands and forests) as the primary cause of their demise.[95]

Protracted volcanism, asteroid impact, and large-scale cooling

(Snowball Earth) have been the leading causes of global mass extinctions, and these catastrophic episodes proceeded at speeds ranging from instantaneous (a massive asteroid impact followed by a profoundly changed biosphere) to geologically rapid. Most notably, precise dating of five volcanic ash beds in Meishan (China) dating from the Permian-Triassic boundary made it possible to constrain the timing of the world's most severe extinction event caused by massive volcanic eruptions in Siberia.[96] The extinction took place between 251.941 million and 251.880 million years ago, which means that its duration of about 60,000 years is equal to only about 0.1 percent of the Permian period that lasted 47 million years. In contrast, the duration of the K-T extinction event remains uncertain, with spans ranging from just two years of global darkness following the impact of the asteroid to hundreds of thousands of years when assuming that the extinction was caused by massive lava flows in India.[97]

After adding the cumulative effect of ever-present gradual environmental changes, competing species, and non-global catastrophes, it is not surprising that perhaps as many as 99 percent of all marine species who ever lived are extinct.[98] Whatever the actual exact (but always very high) share may be, today's species diversity reflects (as it

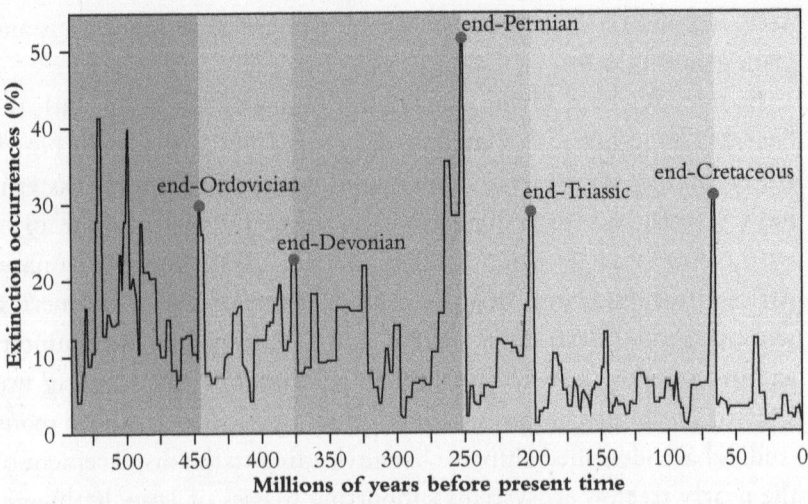

Geologically speaking, extinctions are always with us, but the onset of their spikes and their speed remain unpredictable.

has always done) only a slim excess of speciation over extinction, and that means that good estimates of past extinction rates would give us good approximations of speciation rates. Of course, the inherent incompleteness of the fossil record (above all due to the poor preservation of soft bodies and the destruction of rocks by geotectonic and geomorphic processes) means that such estimates will always remain uncertain.

But thanks to the accumulation of genomic data, it has become possible to reconstruct a global time-tree of life based on nearly 2,300 studies of more than 50,000 species, and to establish rather narrowly bounded speciation speeds.[99] This synthesis shows a remarkable consistency of speciation speeds (dominated by random events) among plants and animals, clustering on the order of 2 million years for complete divergence. Closer looks show average speciation speeds (with all times rounded) of 2.1 (1.7–2.6) million years for vertebrates, 2.2 (1.6–3) million years for arthropods, and 2.7 (2.4–3.6) million years for plants. The overall range of 1.7–3.6 million years per speciation translates (rounded) into 0.3–0.6 speciations per million years.

Another large-scale study that included 25,864 species showed a substantial variation of speciation rates ranging from 0.02 to 1.54 events per lineage per million years, as well as a strong negative relationship between the mean rates of both speciation and extinction and the age of the most recent common ancestor of a group: despite differences in taxonomic diversity, and environmental setting, younger clades (groups of organisms that evolved from a common ancestor) are both speciating and going extinct faster than older groups.[100] Matoniaceae (ancient fan-shaped ferns that have been around for nearly 250 million years and that grow only in Indonesia) was the family with the lowest speciation rate; Lobelioideae (profusely blooming annual plants) was the subfamily with the fastest rate of divergence (about seven times faster).

The formation of species is an inherently complex process, but one simple factor—time—explains a substantial amount of this variation: younger groups seem to accumulate diversity much faster than older groups. But overall, speciation is a slowly unfolding event whose speed is measured best on geological/evolutionary clocks marked in millions of years. The other two key recent findings are that species diversity has been mostly expanding (both overall, and in many smaller groups of species), and the rate of diversification has been mostly constant across the evolutionary span. Genetic drift (the changing frequency of existing gene variants due to random chance) is the leading cause of speciation, and adaptation to changing environments—including shifts in climate, exploitable resources, and habitats—is its other important driver.[101]

The speed of speciation is a field of inquiry where some long-held conclusions have been overturned by recent findings and where a great deal of uncertainty still prevails. Perhaps the best example in the first category is the link between the speed of speciation and latitude. Ever since Alexander von Humboldt and Charles-Marie de La Condamine trudged through the tropical rainforests of South America in the early 1800s, the notion that the richness of species increases as we move toward the equator has been one of the fundamental tenets of modern life science.[102] This higher rate of speciation is best explained by higher energy supply, which speeds up evolution due to shorter generations and higher mutation rates (the evolutionary speed hypothesis).[103]

Some speciation studies of animals (mammals, amphibians, insects) support this conclusion while others do not, but a recent examination of plants—organisms with an obvious latitudinal diversity gradient (just compare the profusion of species in a small patch of tropical rainforest with the paucity of plant species in an equally large patch of tundra)—including more than 60,000 angiosperms (flowering plants), found that tropical plant species have smaller mean speciation rates than temperate ones, with respective averages of 0.65 and 0.73 species per million years, and that the absolute median latitude occupied by individual species is also positively correlated with a higher speciation rate.[104] And a recent theoretical study concluded that a faster evolutionary rate can lower both the abundance of newly formed species and long-term biodiversity. This means that a high evolutionary rate may not be a precondition of having ecosystems with many species.[105]

Obviously, given the millions of species and hundreds of millions of years of post-Cambrian evolution, speciation speeds have ranged widely, with the time needed for the formation of distinct species ranging from less than 10,000 to many millions of years. The evolution of two species of European flounders (an oval-shaped flatfish) living in the Baltic Sea is an excellent example of fast speciation. After the sea became connected to the North Sea (about 8,500 years ago), these two species of flounder diverged quite rapidly. They have different reproductive strategies (one spawns in the open sea, the other one near the bottom where salinity is very low) and the speciation process was accomplished within less than 3,000 generations, or approximately 6,000 years—the fastest speciation speed known for any marine vertebrate.[106]

Humans have become notable new agents of ecological speciation. We have certainly accelerated the process for both cultivated and wild plants—mostly due to hybridization, the combination of two distinct and previously separated species. The hybridization of plants producing new edible species for sedentary agricultural societies has been the most obvious result. Every time you eat bread you are eating fermented and baked dough made from the milled seeds of bread wheat (*Triticum aestivum*), a species only as old as the oldest sedentary societies. This species arose no earlier than 9,000 years ago through the hybridization of a domesticated subspecies of *Triticum turgidum* and *Aegilops tauschii*, an annual grass native to large parts of Asia.[107] No other cultivated plant has been more important for feeding the earliest sedentary societies and eventually establishing itself as a true global staple, cultivated on all inhabited continents.

The evolution of different wheat lineages began about 7 million years ago, but the final hybridization of the two species took place after 10,000 years BCE, when bread, barley, and lentils became the

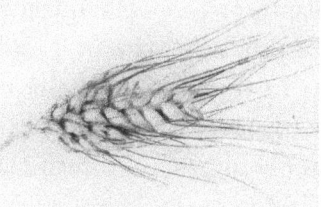

Rapid evolution of a staple: modern wheat (*Triticum aestivum*) and its progenitors, *Triticum turgidum* and *Aegilops tauschii*.

earliest cultivated crops in westernmost Asia. The peanut (*Arachis hypogaea*) is another major cultivated hybrid, combining the chromosomes of *Arachis duranensis* and *Arachis ipaensis*, and so is rapeseed (*Brassica napus*). The plant arose from a spontaneous interspecific hybridization of *Brassica rapa* (field mustard) and *Brassica oleracea*, a species whose cultivars assume diverse shapes ranging from cabbages and kale to broccoli and cauliflower.[108]

The hybridization took place naturally during medieval times; the hybrid's cultivation began about 400 years ago in Europe, and rapeseed was introduced as an oil crop to Asia only during the 1930s. The worldwide cultivation of this crop began to take off in the 1970s, after Baldur Stefansson at the University of Manitoba used selective breeding to develop canola (a contraction of "Canadian oil low acid") rapeseed with very low levels of undesirable erucic acid.[109] In 1980, Canada's area planted to wheat was four times that to rapeseed, but now Canada's canola plantings are about 10 percent larger than those of wheat and yellow is the dominant color when flying across the Prairies in July—as it is in summer across many cultivated landscapes in North America, Europe, and Asia; only soybeans are now globally a bigger oilseed crop than rapeseed.

Among the nearly 100 species of the world's most important crops that have been cultivated during the past 5,000 years, six to eight are new hybrids. That would imply a speciation rate of 10–20 events per million species years. That is exceptionally high: the observed speciation rates of newly hybridized plants are about 50–300 times faster than the expected plant speciation rates.[110] And it also appears that this rate has been accelerating at an exceptional speed.

The global mixing of species due to intentional (the importation of animals and plants) or accidental introduction (the contamination of imported plant products; stowaways in ship holds, and now also on airplanes) has further expanded the opportunities for hybridization between previously isolated species. *Spartina alterniflora*, a perennial saltwater cordgrass of eastern North America, has formed new species in California (by hybridizing with *Spartina foliosa*) and in England, where *Spartina anglica* originated in the 1870s due to the hybridization of the imported American species with the native

Spartina maritima. This new hybrid, spread by human-assisted dispersal, is now present in Europe, Asia, North America, and New Zealand.[111] Another well-studied example of hybridization of wild plant species includes Washington state's *Tragopogon mirus* (remarkable goatsbeard is its common name), a flowering herbaceous plant in the sunflower family that arose from the natural hybridization of *Tragopogon dubius* introduced from Europe and the native *Tragopogon porrifolius.*[112]

England offers other recent examples of the process driven by the human dispersion of species. Eight new species have appeared within the British flora since 1700, and most new hybrid species originating from global plant dispersal have not been detected because of their recent and highly localized appearance or short survival spans, or because they have already become extinct again.[113] As a result, the modern era may have seen a higher speciation rate than at any time since plants colonized the land about half a billion years ago—and that may also be comparable to the ongoing extinction rate. Whatever the actual ratio of these two countervailing processes may be (we can only guess), the unprecedented global dispersal of all organisms is an undeniably powerful and an exceptionally rapid means (on the order of tens to hundreds of years) of new speciation.

At the same time, there can be no doubt that the expansion of human populations and the growth of modern economies initiated a new, major episode of species extinction driven by a combination of the large-scale destruction and degradation of ecosystems (ranging from soil erosion to tropical deforestation), environmental pollution (ranging from heavy metals in farm soils to acid rain), and increasing atmospheric concentrations of greenhouse gases (the leading cause of global warming). An assessment in 2022 by the International Union for the Conservation of Nature and Natural Resources put more than 42,000 species on its Red List of organisms threatened with extinction: shares include 13 percent of birds, 21 percent of reptiles, 27 percent of mammals, 34 percent of coniferous trees, 37 percent of sharks and rays, and 41 percent of amphibians.[114] By 2010, 0.5 percent of known vertebrates had gone extinct sometime since 1500, with about 20 percent of that loss taking place since 1980. As a result, the extinction of vertebrate species has been progressing 24–85 times faster since 1500 than during the great K-T mass extinction, and the post-1980 speed has been even faster, with species loss about 70–300 times faster than during the K-T event.[115] If we add all species that are vulnerable, near-threatened, and those with deficient data, then the speed of the unfolding vertebrate extinction is nearly 9,000 times the magnitude of the K-T mass extinction.

Unlike the previous extinctions, which were driven by infrequent natural (and often catastrophic) events, what has been called the sixth great extinction is the first instance of demise caused by the actions of a single species—but our awareness of this reality also holds a possibility for slowing down, and perhaps even reversing, this rapid trend. And, to close this chapter, how fast was the emergence of our species, and in particular its rise to sapience based on our unprecedented brain size? At what speed did we move first from monkeys to great apes, then to hominoids and finally to *Homo sapiens*?

New World monkeys (including marmosets, tamarins, and howlers) diverged from the mammalian lineage some 43 million years ago, and those animals you can encounter in the forests of Amazonia or Central America have changed little in the intervening time. The separation of Old World monkeys (including macaques, colobuses, and langurs) followed no later than 25 million years ago, and orangutans (three species limited to two Indonesian islands—now endangered in Borneo and critically endangered on Sumatra) were the first diverging branch of great apes (a subfamily of Hominoidea) 15–12.6 million years ago.[116]

Gorillas followed no later than 8.5 million years ago (and have since divided into two species, eastern and western, limited to Africa), and chimpanzee ancestors and the oldest human ancestors (also restricted to tropical Africa) branched out 7.4–6.1 million years ago. Afterwards, the chimpanzee lineage diverged to form two species, bonobos (*Pan paniscus*) and common chimpanzees (*Pan troglodytes*), 890,000–860,000 years ago; and about 420,000 years ago came the divergence of the two common chimpanzee subspecies (*Pan troglodytes troglodytes* and *Pan troglodytes verus*) that entailed no radical physical and behavioral changes.[117]

Recently developed total-evidence dating (combining genome sequencing data with morphological information from both extinct and living species) indicates that the last common ancestor of our genus appeared most likely 3.3 (4.30–2.56) million years ago, and that its brain size was no more than about 420 cubic centimeters, barely bigger than that of a typical chimpanzee.[118] At that point nothing indicated the soaring encephalization (increase in size and complexity of the brain) that was to come. Two species that diverged earlier (*Sahelanthropus tchadensis* and *Ardipithecus ramidus* at, respectively, no later than 6 million years ago and 4.4 million years ago) had brains no larger than those of a chimpanzee.

The slow rise began with *Australopithecus afarensis* (3.85–2.95 million years ago, made famous by the skeleton of "Lucy") whose brain mass, at about 462 grams, was about 20 percent larger than that of an adult chimpanzee. The increase accelerated with *Homo habilis* (about 647 grams; 2.4–1.4 million years ago) and *Homo rudolfensis* (779 grams; 1.9–1.8 million years ago); the Asian *Homo erectus* (1.89–0.11 million years ago) was the first species with a brain mass just above 1,000 grams. With *Homo heidelbergensis* (700,000–200,000 years ago) the mass reached nearly 1,200 grams, and it peaked at about 1,450 grams with *Homo neanderthalensis* (400,000–40,000 years ago); since

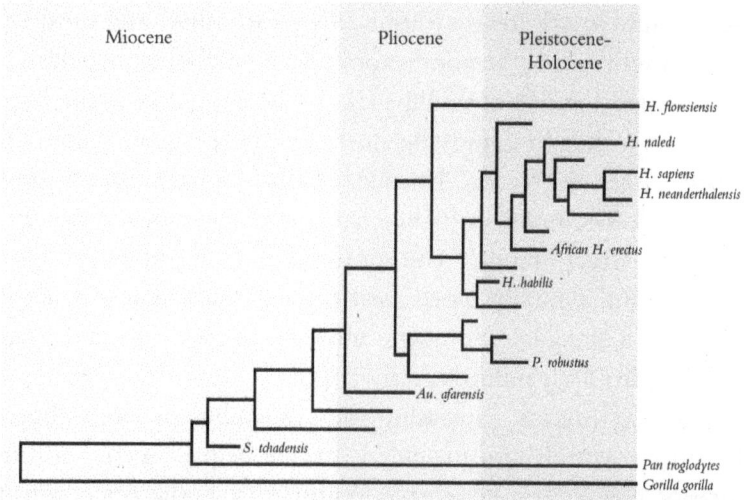

Divergence speeds: from big apes to *Homo sapiens*.

its extinction the less heavy (about 58 kilograms vs. 72 kilograms for adult males) *Homo sapiens* (since about 300,000 years ago) has had the largest hominin brain: total-evidence reconstruction used a mean of 1,397.5 grams, while the larger-scale means from autopsies show weights of 1,336 grams for adult men and 1,198 grams for adult women.[119]

The earliest fossil remains of *Homo sapiens*, discovered in a cave in Jebel Irhoud in Morocco, are dated to 315,000 (±34,000) years ago.[120] With the last common ancestor dated to 3.3 million years ago, this means that brain volumes of species belonging to the *Homo* lineage more than tripled in almost exactly 3 million years, implying an average encephalization speed of 300 cubic centimeters per million years—an unprecedented rate of brain enlargement, especially when compared to the continued stability of brain sizes among other mammalian lineages, be they less than 15-million-year-old great apes or nearly 100-million-year-old rodents, the most successful terrestrial vertebrate order now accounting for some 40 percent of all mammalian species.[121]

What caused this evolutionary surge? The old arboreal theory—the need for stereoscopic vision (the sensing of three-dimensional shapes) and manipulative hands to grasp branches—has been offered as the reason for the high encephalization of primates, but tree squirrels are nimbly arboreal without either capability, and most of the brain growth among hominins took place only after they descended from the trees and became bipedal.[122] The expensive tissue hypothesis posits a shift in deploying nutrients: better-quality diets (more protein, more fat) secured by more skillful foragers made it possible to evolve less complex guts and channel the metabolized energy into producing larger brains.[123] But some extant chimpanzee groups hunt and consume annually more meat per capita relative to their body weight than many later foragers, and their lineage has not shown any notable shifts for 5 million years.[124]

Cooking—that is, increasing the availability of easily digestible nutrients—would have provided further help to secure additional nutrition for energy-expensive brains (at rest, the brain consumes about 20 percent of basal metabolic energy, although it accounts for less than 2 percent of average body mass)—but the rising encephalization is described by a linear function that is independent of the archeological evidence of fire control.[125] In addition, the periods of rapid hominin brain growth do not appear to be associated with adaptive changes caused by environmental unpredictability or by long-term paleoclimatic shifts, including cooler temperatures (the Pleistocene Ice Age), sea-level changes, or increased aridity.[126]

That the high degree of sociality among hominins had its role seems obvious, but primates are not the only mammals living in groups with complex social arrangements. A recent review looked at all major hypotheses offered to explain that momentous change, and evaluated evidence for and against them but did not provide any firm conclusion, while another re-examination of pattern and process in hominin brain evolution also made it clear that this was not a case of consistent directional selection for larger brains, but rather a complicated development that involved such a selection as well as periods of stasis and drift, and even retreat.[127]

All of this is highly unsatisfactory. Achievements and impacts of our species would have been impossible without that remarkable speed of brain growth that began very slowly with the earliest hominins (starting with *Australopithecus*) but by about 2.5 million years ago entered an impressively accelerating phase before reaching its limit— and we would wish to have a far clearer understanding of how and why this encephalization spurt took place. The enhanced provision of nutrients and energy had to have been its foundation, and adaptation to new environments and a higher degree of socialization surely made some difference. But even though the fossil record now at our disposal is far richer (more granular) than it was half a century ago, and even though new techniques of genomic analysis have enabled a more accurate understanding of evolutionary lineages, the great acceleration of hominin brain growth still waits for a clear, convincing explanation.

2. Organisms

Life's strategies and capabilities

A species, the basic unit of scientific classification of living organisms, is defined as a reproductively isolated group of individuals capable of producing fertile offspring and hence propagating its existence. After about 250 years of search, description, and classification—the time elapsed between the first systematic ordering of organisms by Carl Linnaeus in 1735 and the routine deployment of genomic studies before the end of the 20th century—we have identified more than 1.2 million species, ranging from short-lived unicellular bacteria to structurally and metabolically complicated long-lived vertebrates, and we also believe that the actual grand total is significantly higher, most likely close to 9 million eukaryotic (excluding bacteria) organisms.[1] This enormous variety of life means that organisms reproduce and grow at very different speeds, have evolved very different life-cycle strategies, and deploy a remarkable range of mobilities to maximize their chances of survival and propagation.

In this chapter I will first look at the life cycles of organisms—that is, at how fast the candle of life burns: how different organisms reproduce, how rapid or how protracted are their generation times (how fast they grow and reach the age of reproduction), and how long they can expect to live.[2] Then I will turn to examining the speeds of their movement. There are many kinds of locomotion—in water (ciliary, flagellar, pectoral fin, caudal fin, fluke and undulatory swimming, and the jet-based fluid propulsion of squid and cuttlefish), on land (walking, different running gaits, jumping, climbing), and in the air (flapping and gliding flight, suspensory locomotion by swinging in canopies).[3] Speeds range from sluggish (but in relative terms surprisingly fast) bacterial motions on solid surfaces (by twitching, gliding, or sliding) to velocities (particularly in bird-of-prey dives) like those of some modern machines powered by high-performance engines.

I will make some basic observations regarding the life cycles of uni-cellular organisms (bacteria and archaea) and insects (by far the most abundant invertebrates in the biosphere), but most of the attention will be given to vertebrates. Except for the rocks deep underground or under the sea bottom, where only uniquely adapted unicellular organisms survive, these complex organisms inhabit—or repeatedly move through—every niche of the biosphere. Twice a year, bar-headed geese fly over the Himalayas at altitudes close to those of cruising jetliners—where flight demands 20 times more oxygen than when resting, yet the concentration of the gas is less than 8 percent of that at

Extreme locomotion: bar-headed geese crossing the Himalayas, flying above 8,000 meters, and a species of *Pseudoliparis* snailfish swimming at 8,336 meters below sea level in the Izu-Ogasawara Trench off Japan.

sea level.[4] At the biosphere's opposite extreme, there are fishes thriving at depths of up to 8,400 meters below sea level along the Mariana Trench, where the pressure is 800 times that at a depth of 10 meters.[5]

Moreover, vertebrates have combined this ability to survive in an enormous range of habitats with longer life spans and with the mastery of many forms of locomotion. Speeds of reptilian, avian, fish, and mammalian movements will be examined in some detail (looking both at the means and reliably recorded extremes), and the chapter will close by looking at human speeds, particularly walking and running, the two capabilities that have contributed perhaps as much to the evolution of our sapience as has the growth of our extraordinarily large brains.

Life's endless loop: reproduction, growth, longevity

Life starts with reproduction, and single-celled bacteria, the world's most abundant life forms, propagate asexually. They do so mostly by binary fission, doubling their size, and splitting. This cell division consists of the following discrete steps: a cell copies its genetic information (DNA replication), moves the copied genetic information aside (DNA segregation), and then uses special proteins to cleave the cytoplasm (the gelatinous liquid filling the cell) in two and synthesize a new cell wall. Calculations done by Jeremy England at MIT suggest that bacterial division is very efficient. When the heat created during a 20-minute division—involving the formation of new DNA and proteins, as well as cellular rearrangement—by *Escherichia coli* (a very common bacterial species that also inhabits our bodies) is compared with the theoretical minimum amount of heat required by these reactions and activities, the two rates are very similar, indicating that bacterial division is not only rapid but also operates close to its maximum efficiency.[6]

Variations of this process include growing to more than twice the size and then undergoing multiple divisions, and reproduction. While we have many estimates of bacterial doubling times in the laboratory, measuring the speed in the natural environment is nearly

impossible. Perhaps the best way to derive these values is by comparing the annual accumulations of mutations in the wild and the rate
of mutations per generation in the laboratory. Doubling times for
all bacterial species in the wild are much longer than in the laboratory: for *Salmonella enterica* (the common cause of gastroenteritis) the
respective times are 25 hours vs. 30 minutes; for *Escherichia coli*, 15
hours vs. 20 minutes; for *Staphylococcus aureus* (a common cause of
skin and soft tissue infections) nearly 2 hours vs. 24 minutes; but for
Vibrio cholerae (the cause of diarrheal cholera) the difference is much
smaller, 66 minutes vs. 40 minutes.[7]

The life expectancies of most free-living bacteria commonly
encountered in nature and on and inside living organisms are thus
on the order of hours to tens of hours, but very different bacterial
species live deep under the continents and seas. The volume of this
hidden, deep biosphere is almost twice that of the global ocean, and it
contains about 300 times more carbon than the world's human population. Some bacteria are found up to 5 kilometers below the surface,
and beneath the seafloor of the deepest ocean trenches more than
10 kilometers below the water's surface, and their life cycles unfold
on near-geologic scales: because of extreme energy limitations, deep
Earth microbial biomass turns over on a timescale of hundreds of
thousands of years rather than hundreds of minutes—the ultimate
case of life in an exceedingly slow lane spent mostly in suspended
animation.[8]

Asexual reproduction—including budding, fission, fragmen-
tation, and parthenogenesis (whereby eggs develop without
fertilization)—is dominant among fungi, it is not uncommon among
plants, and it is also found among animals (including ants, starfish,
crayfish, lizards, pythons, and Komodo dragons, the largest living
species of lizard found only on five Indonesian islands). But most
insects and vertebrates propagate by the exchange of genetic material, and among these sexually reproducing organisms the first stage
of species propagation—copulation—is often preceded by courting
rituals. Some of these are simple, including insect females stroked
by the males' antennae or legs, insects and birds displaying their
wings or fluttering, and mammals grooming each other or nudging

and wrestling. Others are extremely elaborate, none more so than
the mating rituals practiced by bowerbirds and the birds of paradise
whose males display themselves (often by vigorous dancing) in front
of and inside complicated bowers built from twigs and decorated
with attractive colorful objects.[9]

Semelparous species mate just once per lifetime, followed by death.
Many spiders, insects (some species of butterfly, mayfly, cicada),
Pacific salmon (swimming up freshwater streams to reproduce and
die in the same place they were conceived), and even some mammals
(dasyurids, small marsupials living in Australia and New Guinea) are
semelparous.[10] In contrast, iteroparous species (all birds, most fish,
and nearly all mammals) reproduce repeatedly. The two reproduction strategies are also known as r selection and K selection, using the
letters borrowed from growth-curve terminology where r is the rate
of increase and K is the maximum size.[11]

Insects are the most common r-selected species: whenever conditions are right (after rain, in dead wood, in rotting flesh) they produce
rapidly large numbers of tiny offspring that mature rapidly without any parental care, attributes that make them bothersome and
destructive pests. Among mammals, rodents are the pre-eminent r-
selectionists. The offspring of a single breeding pair of brown rats
(*Rattus norvegicus*) may expand to more than 1,000 individuals in a
year. A fast life means limited longevity (most rats do not make it past
a year), and only parasitic r-selected species, provided with a reliable
supply of nutrients, are long-lived (pork tapeworms can be up to 25
years old).

K-selected species (from antelopes to whales) are slow to reproduce and their single offspring (rarely two; among cattle this happens
once in 200 births) needs prolonged maternal feeding. They are slow
to mature and, if healthy, can live for decades. African elephants
(*Loxodonta africana*) are the perfect example of this slow reproduc-
tion/growth strategy: they reproduce only at 10–12 years of age,
their pregnancy lasts 22 months, a newborn is cared for not only by
its mother but also by females of an extended family, and they can
live commonly for more than half a century.[12]

As for the mating itself, its speed can be extremely fast (cloacal contact in many bird species can last less than a second), while the mating embrace (amplexus) in frogs can last for days, even weeks—but, obviously, all but a tiny share of that long period is inactive: eventually a female expels eggs, they are externally fertilized by a male, and then placed (or hidden) to minimize their exposure to predation. Repeated, and prolonged, mating is common among insects (the southern green stink bug copulates for up to 10 days, the fire bug up to 7.5 days), and there is evidence that repeated mating may increase fecundity in nearly all insects.[13] But copulation of the yellow fever mosquito (*Aedes aegypti*) may be as brief as 13 seconds, and fire bugs may copulate for only a few minutes.[14]

The speed of copulation varies enormously among mammals, but it has a significant but unexpected relationship with body mass. Because copulation exposes animals to a higher risk of predation, and because it claims time and energy that could be used for foraging, it might be expected that smaller mammals (being a more vulnerable prey and needing to feed more frequently than the larger ones) should have faster coitus than the larger animals. Yet an examination of more than 100 mammalian species found that the speed of copulation is negatively associated with body size. For rodents and marsupials (body mass 100–250 grams) copulation can last tens of minutes, or even more than an hour; for large ungulates, dolphins, and whales it can be over in a few minutes or just 10 seconds.[15] The best explanation for this counterintuitive relationship is that small mammals find it easier to maintain copulatory positions than do larger animals.

What comes after depends on the animal phylum. Arthropods (arachnids, crustaceans, and insects) are the largest phylum, with insects accounting for about three-quarters of all identified species. These organisms are characterized by complex life cycles. Ants, bees, beetles, butterflies, moths, and wasps have a four-staged life cycle (egg, larva, pupa, adult) that includes complete metamorphosis, as pupae become adults that look completely different. Crickets, dragonflies, earwigs, and grasshoppers have three stages of life (egg, nymph, adult) and undergo incomplete metamorphosis, with the hatched young ones looking like smaller, wingless adults. Similarly, most marine animals have complex life cycles (from gamete to embryo, then to larva, juvenile, and adult), with at least one of them being mobile.[16]

Gestation periods in mammals are bounded by the extreme values for African elephants (645 days) and tiny Virginia opossums (12 days). Other well-known mammals with gestation periods longer than 400 days include whales, giraffes, and rhinos; pregnancies in donkeys, camels, and zebras last more than a year; for the two most famous large carnivores (tigers and lions), gestation is 90–120 days; while cow pregnancies (279–292 days) are identical to a normal human gestation period (280 days). Mammalian species differ greatly both in their immediate post-delivery capabilities and in the speed with which they reach physical and reproductive maturity.

Human newborns belong to a large class of helpless neonates that require prolonged care before being able to walk (an ability acquired by some at 8 months, and by others only after 18 months), then even longer spans before they can perform basic physical chores and master required behavioral skills, and more than a decade before they can reproduce (the age of menarche has been falling, and it may be now as low as 10 years).[17] In contrast, newly born antelopes (and camels) can stand up within minutes after birth, some species can run within a couple of hours, and many larger rodents reach sexual maturity in less than two months.

Postnatal growth has been studied extensively both in wild species and in domesticated birds and mammals.[18] As expected, the growth rates of small neonates are faster than those of larger species (a 100-gram kitten grows nearly 10 times faster than a 10-kilogram fawn), and they are nearly five times faster in birds than in mammals. Also as expected, warm-blooded animals (endotherms) grow faster than the cold-blooded ones (ectotherms), and the fastest assumed growth of the largest dinosaurs (about 90 kg/day) appears to be comparable to the fastest growth of the largest extant mammal, the blue whale (*Balaenoptera musculus*) at 90 kg/day.[19] In most animals, growth ceases at maturity (determinate growth), while some endotherms (including Asian elephants) and many ectotherms (including tortoises and turtles) experience, after reaching maturity, slower indeterminate growth continuing throughout life.

The postnatal growth of all vertebrate species conforms, often very closely, to one of several forms of S-shaped (sigmoid) curves.[20] Early growth gains are often linear, and then comes an exponential rise leading to the inflexion point (maximum growth rate), followed by slower rates of increase as the growth reaches its limit. Endotherms grow much faster than ectotherms, and animals with the same adult body weights have widely different growth rates: for example, a rabbit will grow more than 10 times faster than a cod.[21] Daily body mass gain maxima for terrestrial animals range from 2 kilograms for rhinos and 0.4 kilograms for elephants to 14 grams for chimpanzees and a mere 10 milligrams for tiny geckos. For comparison, during the first month of their life, human babies gain typically 20 g/day.[22]

The course of human growth differs substantially from that of other placental mammals.[23] The speed of human growth peaks during gestation and then decelerates during infancy, the time of maximal growth for all other placental mammals. The speed of growth (weight and height) is a major indicator of healthy infancy and childhood. During the 20th century we developed detailed weight-for-age and length-for-age (as well as weight-for-length and weight-for-height) standards (national and international), and convenient velocity charts make it easy to ascertain healthy growth or to spot any deficits. These charts show that human growth differs not only from the gains among similarly massive mammals, but also from the weight and height gains of chimpanzees.

Standards adopted for universal use by the World Health Organization begin with average birth weights of 3.4 kilograms for boys and 3.2 for girls. Linear growth during the first two months adds 1.8–2 kilograms, is followed by four to five months of accelerated growth, and then by a slight deceleration with one-year-old infants averaging about 9.6 kilograms for boys and 9 kilograms for girls.[24] This is followed by a period of linear growth and then, once again, acceleration lasting until 14–15 years old, when boys in affluent countries are gaining 5 kg /year and girls about 3 kg/year as they near their adult body mass and height.

This complex course of growth cannot be expressed by basic growth curves that work well for tracing the growth of other vertebrates. As might be expected, specific populations deviate from these gender-specific universal patterns and show non-negligible deviations from the WHO's global growth standards. For example, between the ages of 6 and 10, Chinese boys are significantly heavier than the WHO standard, both boys and girls are significantly shorter when older, and girls of all ages have a much lower body mass index. Moreover, despite China's enormous economic progress, in 2002 about 17 percent of all children were stunted (impaired growth making them short for their age), and a 2010 survey found that about 10 percent of children younger than five years were stunted.[25]

During the 20th century, a higher standard of living and better nutrition eliminated stunting in many nations and resulted in some impressive gains in average height. South Korean women experienced the fastest growth, adding 20.2 centimeters on average, while Iranian men topped the male gains with 16.5 centimeters. American males gained only about 10 centimeters (a millimeter per year), while Chinese gains were briefly reversed by the world's largest famine (1959–1961); even so, since 1950 Chinese males have gained about 1.3 centimeters and Chinese women about 1.1 centimeters per decade.[26] But, remarkably, the height differential between the tallest (including the Netherlands, Belgium, and Estonia) and the shortest (Timor-Leste, Yemen, Laos) male populations increased during the 20th century to just over 20 centimeters.[27]

The last notable difference between the speed of human and other mammalian growth is the timing of sexual maturity: in mammals it occurs soon after weaning; in humans it is postponed for more than a decade and is accompanied by an adolescent growth in stature. As already noted, the age of first menstruation has been declining and it is now as low as 10 years, but male sexual maturity is reached typically between 12.5 and 14 years.[28] And while in many preindustrial societies reproduction began, as with other mammals, shortly after puberty, since then the marriage age and the age of the first (and increasingly only) pregnancy have been rising, extending generation time past the mid-20s. Generation time is the best indicator of the speed with which organisms can propagate: it is the interval between the birth of an organism and the birth of its offspring, and the span varies widely.

At one extreme we have tens of minutes for common species of bacteria grown under optimum laboratory conditions; at the other end we have slowly reproducing primates whose generation time exceeds two decades. We have reliable information on generation times (as well as on maximum longevity) for 5,427 extant species of mammals.[29] Both African elephants and chimpanzees have a generation time of about 25 years; the blue whale, the world's largest mammal, reproduces at the age of 31 years. Short generation times combined with the ability to live in proximity with humans and to benefit from discarded edible waste make almost impossible-to-control pests.

Black rats (*Rattus rattus*) have an average generation time of 143 days, and brown rats (*Rattus norvegicus*) just 91 days (a new litter of up to 12 offspring every three months)—a very fast rate of reproduction when considering that its adult body weight (up to half a kilogram) is as much as 25 times that of a house mouse (less than 20 grams), whose generation time is 63–77 days. In contrast, generation times for great apes differ little from those of our species. Gorillas have average female and male parental ages of 18.2 and 20.4 years, and the study of eight wild chimpanzee communities in West and East Africa indicates the average age of parents of 24.6 years (females just over 25, males a bit above 24 years) and community extremes of 22.5 and 28.9 years.[30] The largest data set available for humans (based on 360 societies) ended up with means of 27.3 for women and 30.8 years for men.[31]

The opposite trend has created modern mass-scale production of pig and chicken meat. The combination of fast growth, relatively rapid reproduction rates, and substantial body mass (along with a propensity for being tamed) explains the choice of domesticated animals grown for meat.[32] Small mammals (rabbits, guinea pigs) grow fast but yield little meat; cattle growth is much slower but meat yield is two orders of magnitude larger. Pigs are ideal meat-producing mammals, and modern breeding and optimized feeding in confinement has combined to produce unprecedented rates of growth. They convert feed to meat much more efficiently than cattle, and a comparison with humans shows the exceptional speed of their postnatal growth. While infants average 3.6 kilograms at birth (about 5 percent of adult mass), newborn pigs weigh only about 1.5 kilograms (less than 2 percent of adult mass), but then they double their weight in a single week, more than quintuple it two weeks later, and are weaned after 25 days.[33]

Young pigs gain 500 g/day, mature pigs gain 800 g/day, and modern lean-meat breeds reach a slaughter weight of 90–130 kilograms just 100–160 days after weaning, or less than six months after birth, at the time when healthy infant boys still weigh only 8–9 kilograms. Production of chicken has seen even greater advances due to modern breeding and feeding in confinement. While free-running birds weighing just over 1 kilogram are slaughtered at the age of 4–5 months, modern broilers need less than 50 days of feeding in confinement to reach nearly 3 kilograms of slaughter weight, and less than a third of the feed per unit of meat than modern pigs. During the late 1950s, a fully grown broiler (after nearly two months of feeding) weighed only about 900 grams; two decades later its weight had doubled, and 25 years later it had more than doubled again!

The combination of this speed and weight of meat production explains why chicken has become the dominant meat in all modern societies and, conversely, why beef (whose production cannot be

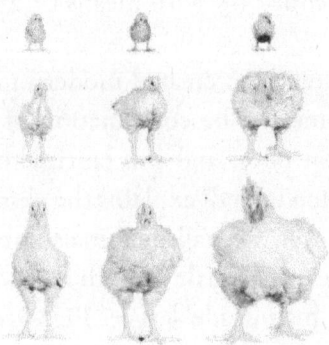

Rapid growth produces a mini monster: broiler front views in 1957, 1978, and 2005, at hatching and after 28 and 56 days.

sped up due to cattle's inherently longer gestation and growth to slaughter weight) has been in retreat.[34]

And one more aside concerning the rate of growth. A fascinating analysis by an international team of biologists has answered one of the most important evolutionary questions: "How fast can a mammal evolve from the size of a mouse to the size of an elephant?"[35] The speed of these radical transformations will determine the speed of recovery from mass extinctions as well as the speed with which new, larger, and more mobile species occupy new environments. Computations considering mammalian body mass changes during the last 70 million years (both in oceans and on land) indicate that for terrestrial mammal mass it took at least 1.6 million generations to increase 100-fold, 5.1 million generations to grow 1,000-fold, and 10 million generations to get 5,000 times larger. Analogical values for whales were only 1.1 million, 3 million, and 5 million generations, and these shorter time spans are best explained by reduced mechanical constraints (buoyancy). But the process is faster in the opposite direction: populations that become isolated on islands will diminish in size, and this decrease is more than 10 times faster than the rate of evolutionary increase. Among the most stunning examples of this insular dwarfism are skeletons of extinct dwarf elephants that lived in Sicily and Malta and whose males were only a meter tall.

Finally, some comments about how long iteroparous organisms (those having many reproductive events during their lives) live beyond their generation times—that is, about life expectancies. They can be very short for invertebrates: even when leaving aside microscopic organisms and looking at common insects, many fruit flies (*Drosophila*) have only a month to live, some small fish only two months, and the shortest mammalian life expectancy, about half a year, is that of the giant Sunda rat.[36] On a species level, the continuation of life can be secured by the rapid reproduction of short-lived individuals or by extending the life of slowly reproducing individual organisms.

Bacteria and protozoa follow the first strategy, and can reach exceptional longevities, some (as already noted) by resorting to suspended animation deep underground, others by having incredible survival capabilities.[37] Tardigrades (water "bears")—tiny (from just 50 micrometers to about 1.2 millimeters) invertebrates that are related both to arthropods and to nematodes, and live on lichens, mosses, algae, and in sand—can turn themselves into cryptobiotic barrel-shaped forms (called tuns) that can survive prolonged boiling in water and freezing to temperatures close to absolute zero.[38]

Multicellularity and sexual reproduction have foreclosed such options, but some animal orders (including whales, tortoises, and petrels) have appreciably longer life expectancies, and within each taxonomic group there are clear longevity outliers. Species that live far longer than the mean survival of animals in their order include the bowhead whale among Cetacea, the Galápagos tortoise among Testudines, and the olm, a cave-dwelling salamander, among Urodela. Looking at the possible environmental parameters and life-history traits that might explain (even if only partially) these exceptional longevities results in a wide range of possible causes.[39] The absence of natural predators is perhaps the common factor: it applies to several whale species, Galápagos tortoises, blind cave-dwelling salamanders, and Andean condors. Living in cold and deep waters or inaccessible places, late maturation, and herbivory (hence no risk of injury in pursuing prey) are other common factors.

Our species, too, is an outlier among primates whose average longevity is mostly between 25 and 30 years, while means in affluent countries are now more than three times longer, and verified individual maxima are nearly five times as long. The highest nationwide means are now nearly 82 years for men and 88 years for women in Japan, and there are 10 other countries (including Australia, Israel, Italy, South Korea, Switzerland, and Sweden) where both averages are above 80 years.[40] The reasons for this extension of human life are clear: better nutrition, better housing, preventive medical care, and advances in treating both chronic and acute diseases have doubled the average human life span during the past two centuries, and this has led to some wishful thinking about potential immortality: aging is perhaps the foremost attribute of conscious existence where speed is most unwelcome! There is no shortage of recent contradictory conclusions in this regard: while there are some clear signs that two centuries of human longevity increase might be approaching the end, other writings argue that amazing breakthroughs await us.

But before we get bedazzled by the promise of vastly extended lives, we should keep the following facts in mind. Mammals have an inverse relationship between their heart rate and life expectancy, and each species appears to have a fixed number of heart beats in a lifetime. Given the range of mammalian body masses and life histories, the total range of heart beats per lifetime is remarkably narrow—within an order of magnitude. Basal energy consumption per body atom per heart beat is identical for all mammals, and the average value of about 10×10^8 heart beats per lifetime suggests that "life span is predetermined by basic energetics of living cells and that the apparent inverse relation between life span and heart rate reflects an epiphenomenon in which heart rate is a marker of metabolic rate."[41] Here speed is an undesirable phenomenon, and non-problematic bradycardia— a slow heartbeat of 40–60 per minute (as opposed to the prevalent range of 60–100 per minute) that does not cause or signify any health complications—would be universally preferable.

Whatever awaits us, humans will remain distinct because our combination of physical and mental attributes sets us apart even from our genetically close relatives, chimpanzees and bonobos. Outwardly the most obvious of these unique capabilities is the mastery of three forms of agile mobility: upright walking, running (ranging from sprinting to what have become incredibly prolonged ultramarathon races), and swimming. But before examining the speeds of human mobility, let's look at animals whose speed (and often also endurance) has surpassed the best human capabilities.

Mobility: swimming, flying, running

Some organisms are subjected to passive locomotion, carried by water or air (jellyfish, spiders), while many parasites (from ticks to hookworms) ride on or inside other animals. Most aquatic animals swim but some also walk (crabs), jump (like a squirting scallop), and crawl (starfish, sea cucumbers). Flapping flight dominates the airborne movements of flying insects, but birds can also glide and soar. On land, movements range from crawling and slithering (worms, mollusks, arthropods, snakes) to walking and running. Unusual modes of locomotion include lizards running on water, snakes gliding, and apes swinging through trees.

Most daily movements are done in search of food, but eluding predators, searching for mates, and building shelters (termite mounts, birds' nests, beaver dams) are also important: quotidian animal activities, and for many species annual movements (most notably for birds and some large ungulates), are dominated by long-distance migrations.[42] The cost of transport—the quotient of expended energy used, and the product of body mass and distance traveled (J/kg·m)—allows us to compare the energy requirements of moving for unrelated species, to get a clear energetic hierarchy of locomotion that is determined primarily by the medium through which the organisms move—that is, by its viscosity and density.

Swimming is metabolically the least expensive (near-neutral buoyancy needs no energy to support bodies), flying costs more, and running is the least efficient mode of animal and human locomotion.[43] All locomotion modes are limited by the metabolic scopes of organisms—the ratio between peak exertion and resting (or basal) metabolic rate. Mammalian scopes are mostly around 10 but go up to 20 for horses and just above 30 for canids. Avian scopes are typically below 20, and so are the scopes for fishes. Maximum speed puts a fundamental constraint on the daily and annual movement of animals—and it would be expected, as with so many other variables, to increase with body mass (M) following a specific power-law relationship (M^b).

But that is not the case, as the largest animals are not the fastest, and the fastest animals (cheetahs, marlins, falcons) are of intermediate size. And animals share two interesting commonalities with airplanes: the biggest animals (elephants, whales), much like the biggest airplanes (wide-body jetliners), move almost constantly, while the fastest animals (predators), much like the fastest airplanes (fighter planes), spend most of their time at rest, and hence their speed average per lifetime is quite low.[44] The most comprehensive analysis of

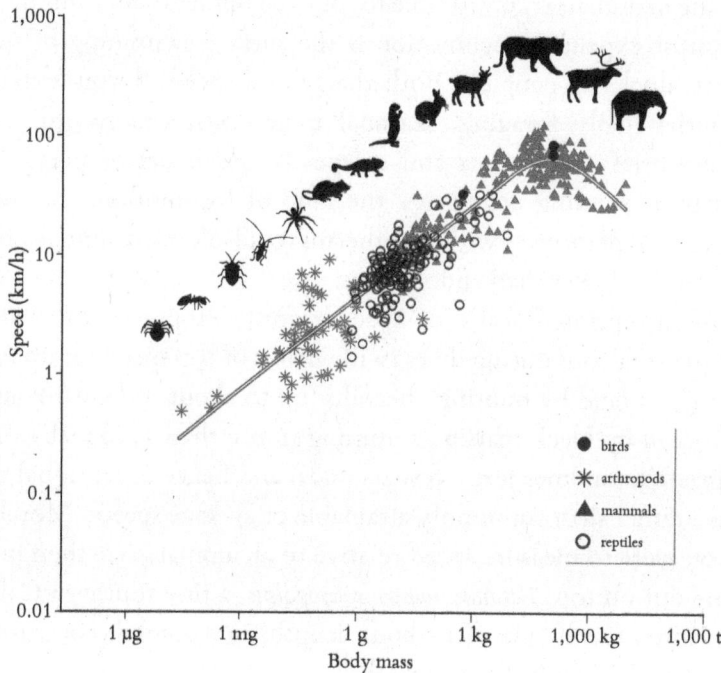

The largest animals are not the fastest: the maximum running speeds of arthropods, reptiles, birds, and mammals.

body masses and speeds—including nearly 500 species (mammals, fish, birds, reptiles, mollusks, and arthropods) with body masses ranging from 3×10^{-8} kilograms to 108,400 kilograms—found that the hump-shaped relationship between mass and speed can best be explained by the fact that maximum acceleration is limited because of restrictions on quickly available energy.[45]

Running and flying animals have an initially similar power-law increase of maximum speed with body mass ($v = M^b$, speed = mass raised to the power of b) but flying animals are nearly six times faster (exponent b is higher). Swimming animals show a faster speed increase with mass: water is 800 times as dense and 60 times as viscous as air, and a bigger body mass brings a greater benefit in gaining speed. We will look at animal locomotion speeds by following the swimming-flying-running order, but I should also mention some notable exceptions: the energy cost of the submerged swimming of marine mammals is comparable to, or even higher than, running, and the most expensive locomotion is the surface swimming of vertebrates, ducks or people.[46] Both the typical speeds during extended activities (daily foraging, seasonal migrations) and record speeds during brief rapid hunts and escapes (as predators or prey) range widely, depending on species, the kind of locomotion, and evolutionary adaptations—with ectotherms (cold-blooded animals) being inherently slower than endotherms.

Often-reprinted scales of speed records—topped in the air by peregrine falcons during dives (with claims of top speeds of up to 389 km/h), on land by hunting cheetahs (up to about 100 km/h), and in the ocean by black marlin swimming (more than 130 km/h)—refer to brief (sometimes just a few seconds) and hence exceptional exertions, rather than commonly attainable or average speeds. Moreover, if you were to measure speed relative to an animal's size then insects come out on top. *Paratarsotomus macropalpis*, a tiny southern Californian mite, travels 322 of its body lengths in a second, compared to just 16 body lengths for a cheetah.[47]

The size (body length) of swimming organisms spans seven orders of magnitude, from a few micrometers to several tens of meters. As just noted, their average speed increases with size faster than with runners or flyers, and among macroscopic species the fastest speeds are attained by vertebrates using lift-based, lateral, or caudal (fins or tails) propulsion. Fresh water density is 1,000 kg/m³ (with only slight temperature changes), while seawater weighs 1,026 kg/m³—but densities of aquatic organisms are higher and adaptations are needed to achieve buoyancy.[48] Most ray-finned fishes, the largest group of living vertebrates, have gas-filled swim bladders; some cephalopods have gas-filled chambers inside shells; many squids contain ammonium-rich body fluids; some sharks come close to seawater's density thanks to having large amounts of squalene (a colorless oil) in their livers, but others (and tunas) support their slightly higher densities by lift on their pectoral fins.

Fish swim by alternate contractions of aerobic muscles that run the length of their bodies in a small red band just under their skin, while their large white anaerobic muscles are used only for brief bursts.

There are, of course, large speed differences among typical swimming modes. For example, in laboratory experiments Atlantic mackerel average less than 0.3 m/s when swimming during the night, 1 m/s during prey capture, and their free-swimming burst speeds are up to about 3 m/s.[49] But there are notable exceptions to this dominant locomotion arrangement: such powerful swimmers as tunas and white sharks have red aerobic muscles deep inside their bodies as well—and, moreover, constant contractions keep those internal muscles significantly (10–20°C) warmer than the surrounding ocean, making them ready either for fast evasion when threatened or for a swift attack.[50]

Published studies put the estimated maximum speed of swimming at about 35 m/s (126 km/h) for sailfish and black marlin, with the great white shark (at 40 km/h) being significantly slower than orcas (55 km/h), while a blue whale reaches no more than 37 km/h.[51] But the fastest swimming speed claims are almost certainly indefensible exaggerations. Perhaps the most fundamental approach to estimating the maximum speed potentially attainable by a fish is to measure its minimum muscle contraction times: after all, fish swimming speeds are limited by tail-beat frequency and minimum muscle contraction times that set theoretical speed maxima. Measurements of the twitch contraction time of anaerobic swimming muscles for sailfish lead to a maximum estimated speed of 8.3 m/s (29.88 km/h), while three other large marine pelagic (open-sea) predatory species—barracuda, little tunny, and dorado (mahi mahi)—have estimated maxima of, respectively, 6.2 m/s, 5.6 m/s, and 4.0 m/s.[52]

This suggests that billfishes, and other fishes using lunate (pointed but not sharply forked) tails for propulsion, cannot exceed swimming speeds of 10–15 m/s, a conclusion reached previously based on three hydrodynamic considerations: the maximal power producible by fish muscles, the limit on the maximal tail-beat frequency, and the onset of cavitation (the formation of bubbles on fish fins and tails that can be painful at high speeds). Cavitation also limits the speed of modern metal ship propellers powered by large internal combustion engines.

Moreover, a study that used acoustic telemetry to collect 175 hours of continuous swimming speed data of three blue marlins (with body mass up to 125 kilograms) showed that they spent most of their time swimming very slowly, just 15–25 cm/s (less than 1 km/h) when near the surface, with very brief speed increases associated with deep dives.[53] And just one more speed range, for the most telegenic of all whales: years of electronic theodolite observations of free-swimming killer whales (orcas) in a British Columbia strait showed rather narrow ranges of speeds: minima of 2.2–2.5 km/h, means of 5–6.1 km/h, and maxima of 7.5–10.8 km/h.[54]

Flying high

Specific energy requirements for flight are determined by the ratio between drag (D) and lift (L), the measure of aerodynamic quality. D/L ratios are around 0.2 for small birds, only 0.07 for seagulls, and at 0.05 albatrosses can do better than jetliners! The reverse value (L/D) is known in English as the glide ratio; in French it has a more evocative term, *finesse* (F). Flying with high *finesse* saves energy and allows for admirably long gliding.[55]

This mode of flying requires no active muscle power, as it converts potential energy to kinetic energy, and it is common among such large birds as albatrosses, condors, eagles, frigatebirds, storks, and vultures. The best gliders, with long slender wings and highly

Gliding with high *finesse*: *Diomedea epomophora*, the Southern royal albatross.

streamlined narrow bodies, have F of 60, losing just 1 meter of altitude for every 60 meters of horizontal progress: from a height of 5 kilometers, they could fly another 300 kilometers before landing!

And even the bulky-looking jumbo Boeing 747 has $F = 15$, which means that if all its four engines fail (a most unlikely possibility), at its cruising altitude of 10 kilometers above sea level it could glide for 150 kilometers to make it to the nearest airport.

Aerodynamic theory predicts that gliding airspeed should scale with bird size and wing shape, but radar tracks of more than 1,300 birds from 12 species did not confirm those relationships, and found instead that average gliding speed unexpectedly converges to a narrow range of about 12.6–15.4 m/s. The best explanation of this discrepancy is that gliding birds adjust their speed to minimize the risk of grounding or switching to a more costly flapping flight.[56] Soaring birds fly without flapping wings, using rising air currents to uplift them. Most seabirds, including the already noted gliders as well as seagulls, pelicans, and terns, are accomplished practitioners of soaring, as are most birds of prey, cranes, herons, and some passerine species (choughs, ravens, wood swallows).

Despite substantial differences in total body mass, wingspan, and *finesse*, two studies based on radar measurements of nearly 140 species of migrating birds whose body mass ranged from tiny passerines (10 grams) to swans (10 kilograms) showed that their airspeeds did not scale as steeply with higher body mass and wing loading as predicted by aerodynamic theory.[57] The first study found that migration speeds for most bird species range between 10 and 15 m/s, with mallards, pigeons, and crows being among a relatively small number of species averaging speeds near or above 20 m/s, while speeds below 10 m/s are rather common among such small birds as pipits, redstarts, wagtails, and warblers. Similarly, the second study found a relatively small range of migration speeds (8–23 m/s), and its authors suggested that some notable evolutionary restrictions work to counteract too-slow and too-fast speeds among, respectively, birds with low and high wing loading (body weight divided by wing area). For comparison, honeybees, those indefatigable insect flyers gathering nectar and pollen, move with speeds mostly between 1 m/s and 2 m/s.[58]

Not surprisingly, migration flights (some longer than 13,000 kilometers in one direction) proceed at optimal speeds that are much lower than brief maximum speeds, with record speed claims (of uneven veracity) going as high as nearly 90 km/h for swans, about 90 km/h for swifts, 127 km/h for the wandering albatross, and 153 km/h for the ascension frigatebird.[59] And while common swifts may not be the fastest fliers, new research (equipping them with micro data loggers and accelerometers) has shown that they are unsurpassed endurance fliers, remaining airborne for more than 99 percent of the time during the 10-month non-breeding period they spend migrating between Europe and sub-Saharan Africa, surpassing the endurance of other exceptional long-distance migrants including Arctic terns and great frigatebirds.[60]

In any case, the fastest airspeeds are not reached in flapping or gliding flight but in dives by birds of prey. High-resolution videogrammetry shows how these raptors change their wing shape in attack, holding their wings depressed below their shoulder and reducing the curvature with increased speed.[61] Falcons begin their attacks hundreds of meters above their prey, and accelerate by beating their wings before folding them and diving, with angles ranging from about 15° from horizontal to a nearly vertical direction—turning potential into kinetic energy and reaching top speeds that were

A diving falcon: ultimate avian high-speed aerodynamics.

estimated by observers to range up to 157 m/s, briefly equivalent to about half of the speed of sound![62]

But these claims were disputed, especially as radar measurements show much lower velocities. Thanks to small GPS data loggers, we now have reliable measurements of diving speeds for five African hunting falconry birds: the devices weighed just 1.5–2.5 percent of the birds' body weight and hence had a minimal effect on their performance. There was a strong positive correlation between maximum hunt speed and maximum flight height for the long-wing species, while maximum and mean flight speeds were negatively correlated with wing area for all five species studied. A female peregrine falcon reached the highest speed of 195.97 km/h during a 20-second flight, with a mean speed of nearly 128 km/h, and her maximum acceleration was about 12 m/s^2.[63]

Running fast

Running is the form of locomotion that is practiced by organisms ranging in body mass from tiny arthropods (millipedes, arachnids, insects) and small-size reptiles (geckos) to large birds (ostriches, rheas, cassowaries) and to mammals of all sizes, from the smallest mice to the largest megaherbivores—rhinos, hippos, and elephants. Running is the fastest terrestrial gait, following walking, trotting, and galloping. Each of these gaits has a different footfall pattern and duty factor—that is, the fraction of a cycle for which a given foot is in contact with the ground. Running has a duty factor of less than 0.5, but fast-moving elephants, whose duty factor is as low as 0.37, are not really running because they always keep at least one foot in ground contact.[64]

The number of steps per unit of distance is inversely proportional to body length (smaller creatures must take many more steps to cover the same distance), while the work accomplished for each step is proportional to body mass. As an animal gets larger, its available power goes up faster than the cost of running—allowing large mammals to run much faster (up to an order of magnitude more) than the smallest ones. Moreover, large mammals reuse significant shares of energy spent running, because a part of the kinetic and potential energy deployed to run is temporarily stored as elastic strain in muscles and tendons and redeployed as elastic recoil.[65]

Once again, speed as a function of mass has a hump-shaped progress. An analysis of body masses and maximal running speeds for 100 terrestrial mammals—with typical adult body masses ranging from more than 6,000 kilograms to 16 grams (African elephant to mouse)—found the optimal size for running to be about 120 kilograms, about the size of a mule deer.[66] Up to that point, for all runners (arthropods, birds, reptiles, and mammals), larger mass is associated with higher speed—but, as with fishes and birds, the largest running animals are not the fastest, and for artiodactyls (even-toed ungulates), carnivores, and rodents the maximal running speed is mass-independent. Still, some of the largest mammals, including some megaherbivores whose mass exceeds 1,000 kilograms, are surprisingly fast. Massive, lumpy-looking hippos, who prefer to wallow or swim slowly in water pools, lakes, and rivers can run as fast as 25 km/h (about 70 meters in 10 seconds), a respectable sprinting speed not to be matched by most adults trying to outrun an enraged animal.

Fast speeds are necessary to escape predators, but often the fastest attainable speed is not the best option—or as one recent study (of Australia's northern quolls, small cat-like carnivorous marsupials) rightly puts it in its title: running faster causes disaster.[67] As expected, a higher escape speed increases the probability of crashes when rounding corners or when escaping along a tree branch.[68] Among predators, mass enhances speed up to about 70 kilograms of body mass, but it diminishes turn capacity: swift predators can outrun their prey, but they may not be able to outmaneuver it. This is an important consideration for all terrestrial pursuit predators: larger animals have greater turn radii and hence they must adjust their pursuit speed to match the cornering ability of their prey.[69]

Decades of African wildlife TV broadcasts have made most people aware of the extraordinarily fast sprints of savannah antelopes, but few would correctly name the fastest species of that family: the common tsessebe (*Damaliscus lunatus*) can surpass 90 km/h; springbok come close (88 km/h), and black wildebeest, Thomson's gazelle, impala, and Grant's gazelle can reach 80 km/h.[70] But most people know that a single cheetah can eventually get even the fastest antelope. Is *Acinonyx jubatus* really the fastest animal sprinter? We now have plenty of accurate evidence to assess this claim and to cite irrefutable records rather than dubious estimates.

Obviously, the fastest speeds are attained only fleetingly, during the opening and middle phases of a cheetah's hunting run, which lasts typically less than 20 seconds and rarely longer than 30 seconds, but for decades the only measurements of the cheetah's running speed were done with captive (or semi-tame) animals chasing a lure on a straight course or by analyzing filmed hunts in open habitats during daylight.[71] The published top speeds of up to 29.8 m/s (or 107.3 km/h) were significantly faster than those of greyhounds bred for speed and high pulmonary capacity. A detailed comparison of the musculoskeletal anatomy of cheetah and greyhound limbs explains why greyhounds can do no more than 17 m/s.[72]

These animals are of similar size and of similar gross shape—and, surprisingly, cheetahs have a smaller volume of hip extensor musculature than greyhounds. But they use their extensive back muscles to accelerate, their hindlimb bones are proportionally longer and heavier (making longer strides possible), and their extremely powerful psoas muscle (in the lower lumbar region of the spine, extending through the pelvis to the femur) helps to resist the pitching around the hip that is associated with fast accelerations. No less important is the fact that a cheetah supports 70 percent of its body weight on its hindlimbs when running at 18 m/s, and a greyhound just 62 percent of its body weight. Supporting a higher share of body weight on those limbs reduces the risk of slipping.

Finally, cheetahs also have higher stride frequency (up to four cycles a second) than greyhounds (about 3.5). But heat dissipation limits the exertion: when sprinting at close to 100 km/h, a cheetah's heat production would be more than 60 times greater than at rest, and as most heat (70–90 percent) is stored rather than immediately dissipated, the maximum pursuit distance will be limited by this as the animal comes close to overheating. At rest the animals dissipate heat efficiently (maintaining a constant body temperature of about 40°C, even when the air temperature is 10 degrees higher), but this evaporative heat loss is not enhanced during fast runs.[73] The high energy cost of these exertions is well illustrated by comparing breath frequencies: 16 breaths/min at rest, up to 156 breaths/min after a dash.[74] Obviously, the need for subsequent rest and recovery limits the frequency of such chases.

By far the best evidence of the cheetah's hunting behavior comes from monitoring nearly 400 runs of five radio-collared wild cheetahs in Botswana, equipped with GPS, accelerometers, and other devices to provide highly accurate information about their position, direction, speed, and movement.[75] The fastest run was 25.9 m/s (93.2 km/h), but the average run was only 173 meters (the longest one 559 meters) and the mean top speed of all runs was only 14.9 m/s. At the latter speed the animal's turning radius would be 52 meters, and a 180-degree turn would take six seconds to complete, far too long and much too slow to be of any use in capturing a quick-turning prey. Consequently, most hunting runs proceed only at moderate speeds, and the animals can slow by 4 m/s in a single stride to make tight turns, using their tail for balancing and their ridged footpads and claws to maintain good grip, and hence they can successfully hunt not only in open grasslands but also in densely wooded areas.

Cheetah hunts nearly always involve not just a pursuit but also a series of evasions, and hence hunting dynamics are greatly influenced by the behavior of differently sized prey. A cheetah will first accelerate (up to 7.5 m/s²) to close on its prey (achieving hunting speeds close to 19 m/s), and then reduce its speed to make it easier to execute the rapid turns required to match prey escape tactics: as the team who used GPS and tracking collars to study cheetah hunting runs concluded, "Predator and prey thus pit a fine balance of speed against manoeuvring capability in a race for survival."[76] This has been confirmed in a more general manner by simulations of predator-prey encounters: escape success is determined by both speed and maneuverability, and hence a highly maneuverable slow prey can still escape predation while a poorly maneuverable prey can escape only when outrunning its predators.[77]

Not just how fast, but also how agile: a cheetah closing in on a Grant's gazelle.

Horses: from wild to tame

The horse (*Equus ferus caballus*) is a domesticated odd-toed ungulate that has, for millennia, provided humans with the fastest means of land transportation, and its high-bred specimens continue to delight large numbers of horse-racing fans worldwide with their competitive runs. Their top speeds are attained under very different circumstances than for cheetahs, during races that extend most commonly for 1–2.4 kilometers (5–12 furlongs) and can be timed very accurately. But before looking at the speeds of the fastest tamed quadruped, we should consider how we domesticated these beasts.

Horses were not domesticated during the first wave of the taming of wild animals, which coincided with the domestication of staple grain crops in Western Asia about 10,000 years ago and included cows, sheep, and goats.[78] Archeological evidence shows that the earliest domesticated horse lineage (with documented bridling, milking, and corralling) goes to Central Asia's Botai culture at about 3500 BCE, but that the modern domesticated breeds do not descend from it, and a large-scale analysis of nearly 300 ancient horse genomes pinpointed the Western Eurasian steppe in general, and the lower Volga-Don grasslands in particular, as the homeland of modern domesticated horses.[79]

Starting about 4,000 years ago, these domesticated breeds rapidly expanded across Eurasia and replaced all but a few remaining local wild populations. Domestication involved strong selection for critical locomotor and behavioral adaptations (speed, endurance, a stronger backbone to carry loads, docility) and was accompanied by the rise of equestrian material culture, including the evolution of horse-drawn vehicles, for example spoke-wheeled chariots. Since their domestication four millennia ago, horses have been prominent in many human activities, ranging from travel and fast messaging to playing roles in many religious, political, and social aspects of pre-modern civilizations. Their leading use in warfare endured for most of the 19th century, and large numbers of animals were used and died even during the Second World War.[80]

The historical record finds horses entombed as a part of cultic rituals, from European Iron Age burials with chariots to the massive sacrificial horse pit of hundreds of skeletons belonging to the tomb of Duke Jing of the Chinese state of Qi. They were used in public races, from the imperial charioteers of Rome's Circus Maximus to the bareback relay races of Great Plains native tribes.[81] They were harnessed for working in the field, and later factories, being the most powerful widely used animate prime movers before the invention of steam engines. They were taken overseas on long-distance conquests: Hernán Cortés had 1,300 soldiers and 16 horses for his improbable conquest of the Aztec capital, Tenochtitlan, and Bernal Díaz del Castillo, in his memoirs of armed campaigns, left us a detailed description of 16 of their individual capabilities ("easily turned," "not good for anything," "a great runner").[82]

Horses were bred for stature and endurance, with massive Shires, Clydesdales, Percherons, and Vladimirs taller than 18 hands (183 cm) at the withers.[83] They were valued by emperors and dubious traders and have been sold at great horse fairs from Ireland's Cahirmee (where, contrary to Irish lore, Bonaparte did not buy his mount, Marengo, in 1799) to India's 500-year-old and still-thriving Chetak Festival in Sarangkheda.[84] They were slaughtered during and fatally injured after every great battle—from Carrhae, where the Romans were defeated by the Parthians in 53 BCE, to Austerlitz in 1805 where Napoleon's Grande Armeé defeated Russian and Austrian forces.

And they have been eaten, from an unlikely Kazakh *qarta* (boiled and fried horse rectum) to bright red, raw Japanese *basashi*.[85]

German historian Reinhart Koselleck went so far as to elevate these—in his assessment—neglected protagonists of history to determining factors in epochal division: the pre-horse age that he dated to before about 4000 BCE (we now know a new, later, dating would be needed); the horse age, from 4000 BCE to about 1950 CE, and the post-horse age of the last 70+ years.[86] Another German historian, Ulrich Raulff, looked at the separation of man and horse in the 19th century as the great divide that first diminished and then eliminated the animal's historical significance during the modern era—above all its obvious key role as the provider of affordable and manageable kinetic energy superior to human capabilities. There is no doubt that without this unique animal vector, the history of our civilization would not have followed the course it did.[87]

Horse breeders and riders have accumulated an extensive understanding of the animal's capabilities, resilience, and vulnerabilities. And even now, in a distinctly post-horse age, there is no shortage of often elaborate studies looking at the physiology, energy use, endurance, and running performance of horses—including the use of GPS to measure the speed of running, and treadmills and face masks to investigate oxygen consumption. This means that our understanding of horse locomotion rests on solid observational and experimental foundations, and we can draw many solid conclusions.

Horses have four distinct gaits: walk, trot, canter, and gallop, and they can be trained (such as the famous Viennese white Lipizzaners) to perform other (symmetric and asymmetric) movements.[88]

The four-beat walk of the horse is comparable to the human walking gait and its average speed is a bit less than 2 m/s, or about 6.5–6.9 km/h. Trotting (a two-beat rhythm reminiscent of human jogging) is at least 13 km/h, and cantering (a three-beat gait) is between 16 and 25 km/h. Galloping (four-beat) speeds can be up to twice as high, typically ranging between 40 and 50 km/h (that is, up to about 13.9 m/s). Most horses can sustain such speed for just 1–2 miles (1.6–3.2 kilometers). Galloping strides are 5.7–7.5 meters long; the animal's body is suspended once in each stride, and the leading front and trailing hind limbs support the body longer than the opposite extremities.

A change of lead in the gallop takes place first for the front feet; the running animal's center of mass is near its withers (the ridge between its shoulder bones) with the back relatively rigid. A study of the velocity and stride of galloping Thoroughbred horses found a mean stride length of 7 meters, a mean frequency of 0.43 strides per second, and a speed of 16.63 m/s. Predictably, both speed and stride length decreased, and stride count increased, with race progression, and there was substantial inter-horse variation in stride parameters with speed predicting less than half of it.[89] Fit and healthy horses can maintain a speed of about 50 km/h (13.9 m/s) for 8 kilometers, 30 km/h (8.3 m/s) for 30 kilometers, and their running speed declines slowly: it can still average about 25 km/h (7 m/s) over a distance of 45–50 kilometers. Only the Mongolian wild ass is a better runner among large ungulates.

Horses have been used in six distinct roles. As draft animals in farming (mainly for plowing); to provide stationary power (for pumping water from wells or lifting materials from mines to the surface); in road transportation (pulling carts and wagons); as the fastest means to deliver urgent messages; in mobile warfare (initially without saddles and stirrups, and eventually even with body armor); and in racing.[90]

Military assaults and racing aside, these tasks involved limited speeds, but in the early modern era some animals—well-fed and harnessed to small two-wheel carts or to light buggies and phaetons (open carriages with large wheels and minimal bodies)—could trot and canter easily, and even gallop for short stretches on (rare) good roads. Racing speeds improved after long periods of selective breeding, and we now have large databases of reliably measured maximum speeds in horse racing. The records do not belong to Thoroughbreds (descended from English mares and imported Middle Eastern stallions going back 2–3 centuries) but to American Quarter Horses, whose name alludes to their supremacy in quarter-mile (402-meter) runs.[91] The fastest measured quarter-mile runs for Thoroughbreds and Quarter Horses differ only by a tenth of a second.

Standing vs. running starts make a substantial difference. High-speed cameras show a delay of 0.35 seconds between the gates beginning to

open and then being fully open, and then an additional 0.23 seconds
for the first foreleg to come down in the first stride, losing nearly 0.6
seconds before the horses clear the gate. Comparisons over quarter-
mile runs show Quarter Horses posting the best times and gaining
speed in each segment of runs shorter than 336 meters, while in longer
runs Arabians and Thoroughbreds reached their highest speed during
the middle of the race. A recent comparison of Thoroughbreds and
Quarter Horses shows why the latter run faster.

As with every runner, stride length and stride rate are the two
main factors determining speed. Measurements showed the Thoroughbreds have longer strides (7.4 vs. 5.9 meters) but Quarter Horses
have higher stride rates (2.88 vs. 2.34 strides per second).[92] Analyses of
complete Thoroughbred runs over longer distances (1,300, 1,900, and
2,300 meters) based on the French racing data at Chantilly show that
horses reach peak velocity (close to 19 m/s) less than 300 meters after
the start, and can maintain speeds above 15.15 m/s—with inevitable
speed decreases in bends—almost until the end, when their pace
declines slightly as their anaerobic energy supply gets exhausted.[93]

And the analyses of winning speeds for the US Triple Crown
races (the Kentucky Derby, Preakness Stakes, and Belmont Stakes,
first run in 1875, 1873, and 1867 respectively) have shown that there is
almost no room left for any faster performances. There was no significant correlation between year and winning speed for the first race
between 1949 and 2008, the second one peaked in 1971, and the third
in 1973, with the best-fit lines showing plateau speeds at 16.5 m/s for
the Derby, slightly above it for the Preakness, and slightly below it
for the Belmont.[94]

Clearly, there is an absolute upper limit to speeds in each of these
races, and predictions of absolute maximum running speed are only
marginally faster than the current records.

What can be improved is the running strategy in longer races. The
optimal strategy requires the maintaining of maximal force at the
end of the race, and hence those horses that tend to slow down too
much at the end should run less strenuously at the beginning and
slow down only slightly during the whole race to be able to maintain finishing velocity. The top speeds reached during actual races are

Speed

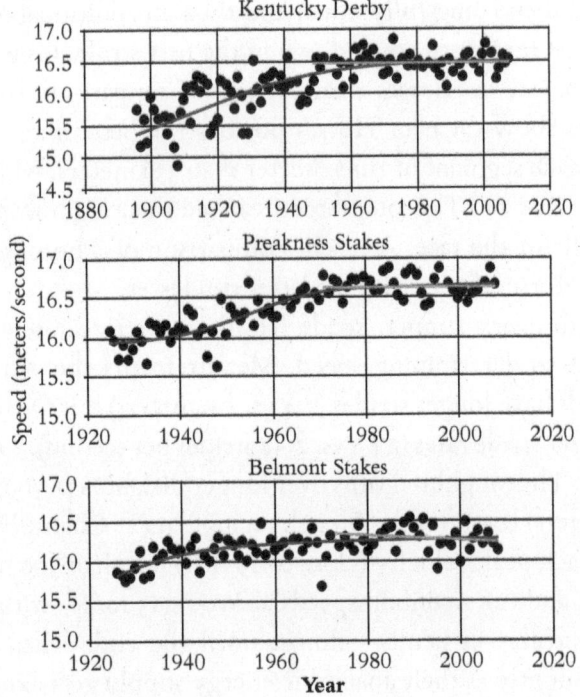

A century of winning speeds in the Triple Crown races: all of them
reached their maximum speed.

60.86 km/h for a one-and-a-half-mile race, and 70.76 km/h for the
quarter-mile run. That record was set in 2008 by Winning Brew, a
two-year-old filly, and highly trained horses can achieve briefly (over
the course of tens of meters) speeds of up to 88 km/h (55 mph, or
24.4 m/s).[95]

Locomotion: How do we move?

Humans lack any radical avian adaptations required for birdlike
flapping flight (powerful breast muscles, long forelimbs to support
wings able to provide sufficient lift and thrust, a superior respiratory
system), and we do not have any extensible patagia (membranes) that
would allow us to glide. The majority of today's humanity cannot
swim, those untrained adults who do swim move no faster than very

slow walking, and even elite swimmers go no faster in brief spurts than the maximum speed of a goldfish. But we are outstanding walkers and admirable runners as we combine endurance with speed. Every year, millions of people run marathons (42.195 kilometers), and increasing numbers of runners have been competing in ultra-long races, with the longest one—the Self-Transcendence 3,100-mile (5,000-kilometer) run across the United States—taking (with short daily sleeps) 40–50 days to finish.[96]

Walking—in physical terms, a pendulum of swinging legs exerting forces on the ground—is an activity for which almost everybody has a good sense of typical performances.

A kilometer in 15 minutes (4 km/h, or 1.1 m/s) is a generally accepted slow-walking speed; 5–6 km/h is a well-paced walk. We can walk at speeds we would prefer to run at (race walking is an Olympic discipline based on that option) and vice versa.[97] Stature (body length) explains the apparent body-size dependency of human walking economy. Most economical walking speeds are dynamically equivalent between smaller and larger individuals, with stride lengths directly proportional to stature, while the energy cost per stride varies by only a small amount.

This close coupling of height, gait mechanics, and the energy cost of walking produces an inverse link between the mass-specific energy costs of walking and stature. This means that people spanning a wide range of ages, statures, and weights expend the same mass-specific energy (J/kg) when walking a horizontal distance equal to their stature.[98] Between about 2 and 2.5 m/s (less for children with shorter legs), walkers voluntarily switch to running. Walking races are a notable exception because a peculiar movement of the hips allows walking (always keeping one foot on the ground) as fast as 4 m/s. As with any physical activity, older people walk slower, and because walking underpins a majority of daily tasks, the measurement of gait speeds is a valuable predictor of overall health: gait speed should be seen as another vital sign alongside the much more commonly measured blood pressure.

The most common standardized test is the timing of a 4-meter walk along a straight path and on a flat surface: the total distance is 9 meters, with the first and last 2.5 meters for acceleration and deceleration. Typical gait speeds in people in their 70s are 1.13 m/s for women and 1.26 m/s for men; during the 80s and 90s the respective speeds decline to 0.94 and 0.97 m/s. Speeds of less than 0.8 m/s are predictive of poor clinical outcomes, and speeds below 0.6 m/s in individuals who are already experiencing serious deterioration confirm diagnoses of continued decline.[99] In contrast, hospital patients in acute care cannot typically move faster than 0.45 m/s; for outpatients the rate goes up to 0.74 m/s.

How do we run so fast?

Running speeds, from casual to the fastest sprints, have more than a three-fold range. Untrained runners can manage 3–4 m/s for 15–30 minutes; the record marathon speed is 5.8 m/s (it was only 4 m/s in 1910); world records in 5-kilometer and 10-kilometer runs were achieved with average speeds of, respectively, 6.4 and 6.2 m/s; and world records for 1,500-, 800-, and 400-meter races translate into average speeds of, respectively, 7.28, 7.92, and 9.29 m/s. The 1975 switch from manual to automatic electronic recording does not make the historical records of the world's fastest race, the 100-meter dash, strictly comparable, but the best time at the first Olympics in 1896 was 12 seconds, and by the end of the 20th century (the 2000 Summer Olympics in Sydney) the time was reduced to 9.87 seconds for men and 11.12 seconds for women.[100]

The fastest runs must combine rapid reaction time (time to respond to the sound of the starting gun), optimal stride length (limited by an individual's size) and frequency (which can be increased by muscle composition and training), and sufficient metabolic power.[101] Faster sprints are made more difficult, as air resistance increases while ground-contact time and muscle-force production decrease. Leg length obviously limits stride length, and while more muscle mass can produce more power (resulting in greater ground-reaction forces), the energy cost of accelerating a heavier mass goes up.[102]

The latest 100-meter sprint world record was set by Jamaican sprinter Usain Bolt at the 2009 IAAF World Championships in Berlin, and detailed analyses of this performance and of the capabilities of the world's 10 fastest men illustrate the magnitude of this achievement. The all-time-high gait characteristics of the world's 10 fastest sprinters (average height of 1.82 meters, and average best time of 9.74 seconds) show a mean stride length of 2.23 meters and mean strike frequency of 4.62 per second.[103] At the time of his 2009 record-setting run, Bolt was significantly taller (1.96 meters) and heavier (90 kilograms) than the other finalists (averages of 79 kilograms and 1.77 meters); in this race his stride (2.47 meters) was nearly 11 percent longer, his strike frequency nearly 7 percent lower, and he made 40.92 steps compared to 44.91 for the group's mean; his mean velocity of 10.44 m/s was 3.5 percent faster and his overall time (9.58 seconds) was 16 milliseconds faster, and in the 70–80 m stretch Bolt reached a speed of 12.34 m/s.[104]

There is a simple way to be awed by this speed: mark the distance of nearly 12.5 meters on level ground (or, if you have a large home, in your living room), and then try to imagine how it could be spanned in a mere second! And while Bolt came close to the perfect combination of sprinting attributes, a truly ideal sprinter could do even better ("better" in this context means a few milliseconds). Bolt's

Still unbroken: Usain Bolt's 2009 record 100-meter run.

longer contact time (0.091 seconds) and lower step frequency resulted in vertical and leg stiffness that were significantly lower than those of Tyson Gay and Asafa Powell, the two men with the second- and third-fastest 100-meter runs (high leg stiffness is necessary for maximum velocity). Moreover, Bolt could have improved his record time without any additional muscular effort.[105]

The needed milliseconds could have come from three gains. First, from improving his only relative weakness, his unremarkable reaction time. A study of more than 1,300 sprinters running between 2003 and 2009 showed that male finalists averaged 142 milliseconds while Bolt's reaction times were, respectively, 166 and 165 milliseconds during the Beijing and London Olympic finals, and 146 milliseconds during his world record Berlin sprint.[106] The other two savings could have come from racing with the maximum permissible wind (2 m/s) and at 1,000 meters above sea level—the maximum altitude permitted for setting world records. That combination could have set a record of 9.45 seconds and an average speed of 10.58 m/s.

Bolt retired in 2017 without ever improving his reaction time or trying to stage a record under the best permissible conditions. Injuries aside, even at the age of 31 he was not past his possible peak, because age-related running speeds do not decline drastically until much later. Their course evolves much as many other human capabilities: rising from low rates (below 3 m/s in young children) to peak performances, for both men and women, between the ages of 27 and 30, but with girls' speeds developing faster than the average running speed of boys.[107]

The subsequent decline is gradual, with maxima only about 10 percent lower by the age of 45 years, and 20 percent lower for males and 25 percent lower for females by the age of 65 years. Only after 70 years of age do speeds decline to levels prevailing among 10-year-old children. For example, men able to run a 10-kilometer race at 75 will do so, at best, at half of their own top speed from earlier in life (their best performance would have been at age 28). These findings are rather encouraging, as typical declines remain relatively modest for three to four decades following the peak. And, of course, prospects are even better for walking: unless restricted by musculoskeletal ailments, healthy adults over 70 years of age can keep walking at a good pace.

Running is energetically demanding as it requires 10–20 times the basal (resting) metabolic power, and the energy cost of human running is higher than in similarly massive mammals, but because of bipedalism and sweating (highly efficient heat dissipation) we have a unique ability to uncouple the energy cost from typical (2–6 m/s) running speeds.[108] Running quadrupeds can breathe only once per locomotor cycle because their thoracic bones and muscles must absorb the impact on the front limbs, but human breath frequency does not have to correspond to stride frequency. Each of us has an optimum walking and an optimum running speed—that is, a speed at which the energy cost of locomotion is distinctly less costly than at other speeds. Graphs of speed vs. energy cost are curvilinear: for walking, the minimal expenditures are at an average speed of 1.3 m/s; for running, at 2.9 m/s for females and 3.7 m/s for males, both significantly correlated with body mass.[109]

As for thermoregulation, adults can evaporate more than half a liter of water per square meter of skin (five times the rate for horses), with peak hourly rates surpassing 2 liters.[110] Another remarkable ability is that while running we can tolerate periods of moderate dehydration and hence can endure prolonged runs even during the hottest part of the day without overheating, a feat that eludes even antelopes, the fastest hot-climate animal runners.[111] A common demonstration of this ability is marathon runners, who do not drink enough to replenish water losses during the race and may take more than a day to completely rehydrate. More than that, marathon runners should avoid excessive water intake in order to prevent hyponatremia—a low sodium level that occurs in a significant fraction of non-elite marathon runners and can be deadly, and happens as water dilutes the sodium they have in their bodies (hence the prevalence of electrolyte sports drinks).[112]

This combination of running and temperature control capabilities made humans uniquely equipped to act as high-temperature predators that could rely on persistence running to chase animals to exhaustion.[113] The topic of persistence running has been the subject of many direct investigations and hypotheses, but after setting many remaining questions, controversies, and critiques aside, there are a few incontrovertible facts. Modern studies merely caught the very end of what must have been a more common way of hunting in the past; such studies have accurately documented the speeds, durations, and success/failure rates in the central Kalahari, with most of the hunts averaging 4.8–6.6 km/h (ranging from just walking to a combination of running and walking), making it clear that persistence hunting was a practical and rewarding option for premodern hunters.[114]

Two feet on the ground

Bipedality is the precondition of these accomplishments, and bipedality has been—along with large brain size, speech, and copious sweating—one of the key evolutionary advances that have made *Homo sapiens* so unique that it has been able, for better and for worse, to emerge as the dominant global species. Bipedality freed our arms and hands for countless practical tasks; it gave us new vistas and allowed us to surpass the mobility of hominids (all modern and extinct great apes). Because our species has become so good at it, walking (and running without falling) seems effortless, even on an uneven surface—but it requires constant and complex adjustments to maintain stability, and we are able to predict the corrective actions needed during the next stance well in advance of touching down our foot.[115]

The best runners have become steadily faster during the past three centuries, for which we have record runs to compare, but in the early 21st century their speeds have come close to the expected asymptotes of the fastest possible performances limited by human capacities. For the 18th and 19th centuries we have analyses of the results of the UK's popular long-distance foot races.[116] Times for 10- and 20-mile runs were, respectively, 10 percent and 15 percent faster during the 20th century than during the 19th century, but those were only about 2 percent faster than in the 18th century.

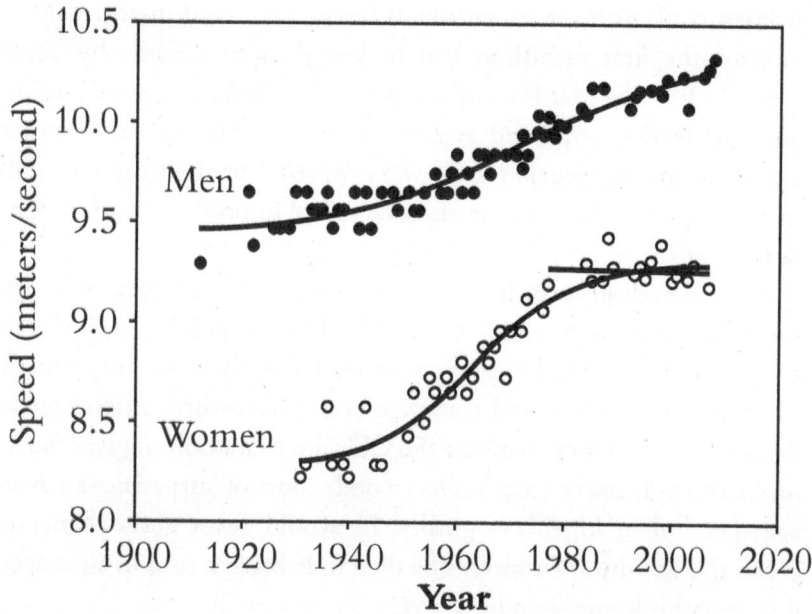

Men have approached the maximum possible speed in the 100-meter race;
women appear to have reached it already.

Analyses of world records for the entire 20th century show significantly higher gains for long distances: record speeds for the 100-meter sprint rose by 8.1 percent for men and 22.9 percent for women.

For 400 meters, the 20th-century gains were 9.7 and 25.7 percent; for 1,500 meters, 12.4 and 10.3 percent; for 10 kilometers, 15.1 and 8.5 percent; and for the marathon, 21.5 and 38.6 percent.[117] When recalculating in terms of average speed gains, the long-term linear increases were 0.11 percent a year for the 100-meter sprint and 4 percent a year for the marathon. Most female records appear to have reached the expected best time, while the speeds for male races are very close to the expected tops both for sprints (10.55 m/s for 100 meters, 10.73 m/s for 200 meters) and for middle distances. When Henry Ryder and his colleagues examined the 20th-century gains in 1976, they concluded that new records could be set for decades to come; Whipp and Ward (1992) predicted that, by the year 2000, women could run the marathon as fast as men; and in 2011, Joyner forecast the first marathon run in less than two hours by about 2021.[118] Only the first forecast came to pass. Marathon speeds, measured by deciles, improved at every level of performance (between 2007 and 2013, the world record was lowered four times), and by 2011 speeds for 100-mile ultramarathon races had improved even faster, by 14 percent.[119]

New marathon records were set in 2014, 2017, and 2022, with the best time (2:01:09, or about 5.8 m/s, by Eliud Kipchoge in Berlin on September 25, 2022) being 134 seconds faster than the 2013 record. But it took only a year and 13 days to break that record, when Kenya's Kelvin Kiptum Cheruiyot ran the Chicago Marathon in just 2:00:35 on October 8, 2023, a mere two seconds short of surpassing an average speed of 21 km/h! Tragically, he would never get a chance to break the two-hour barrier, as he died in February 2024 in a car accident near his hometown in Kenya.

Similarly, new female marathon records were set in 2017, 2019, and 2023, but, although the difference has been shrinking, the fastest time—by Ethiopia's Tigst Assefa (2:11:53), set in Berlin on September 24, 2023—is still 11 minutes and 18 seconds slower than the male record. Marathon speeds have improved at every level of

Twenty-three-year-old Kelvin Kiptum Cheruiyot running the record marathon in Chicago in October 2023.

performance; by the beginning of the second decade of the 21st century, 94 percent of the 100 best men runners were African, while among women athletes that share was 52 percent; the stature, body mass, and body mass index of the top runners had all decreased; and the best running times were achieved in spring and fall when average temperatures are close to optima for long runs.[120]

Ultramarathons (with distances of 50 and 100 kilometers and 50 and 100 miles) have seen even greater relative speed gains. The top performances for 100-mile runs were just below 1,000 minutes in 1998, by 2022 they were below 800 minutes, and by 2022 the record was below 12 hours (11:45:56, or almost 3.8 m/s).[121] And beyond that is the world record of endurance runs: 24- and 48-hour races run, inevitably, at slower speeds. Detailed biomechanical studies of running have demonstrated that higher speeds are the result of applying greater support forces (force/mass) to the ground, not by a more rapid repositioning of legs. While the human body is mechanically complex, ground forces can be understood by looking at just two body parts (the foot and the lower leg that are stopped abruptly by impact, and the rest of the body above the knee that keeps on moving) and three basic stride variables (contact time on the ground, time in the air, and the motion of the ankle or lower limb).[122]

The initial force arises from quickly stopping the lower part of the leg (the shin, ankle, and foot) as the foot strikes the ground. Olympic-level sprinters accomplish this rapid deceleration by forefoot landings. Support forces to the ground (which may exceed the body mass five-fold in sprints) increase with speed, while the time taken to swing the leg into position for the next step does not vary. In addition, foot muscles act as dampers and motors, dissipating and generating mechanical energy: at all running speeds the foot absorbs energy from early to mid-stance, and subsequently it returns part of the absorbed energy in late stance, with the return ratio rising with running speed.[123]

Swimming is inherently more energy-intensive than running: this form of human locomotion is made easier by the body's buoyancy, but the propulsive force of strokes is required to overcome the resistance of the medium (water is about 830 times denser than air) as well as the drag created by the swimmer. Three kinds of drag come into play: friction, pressure, and wave drag. Frictional drag is due to constant collisions of the swimmer's body with the surrounding water molecules, and it increases with higher speed. At higher speeds, water also begins to build up around the swimmer's head, resulting in the pressure difference between the two ends of the swimmer's body—which generates turbulence, an additional source of resistance. And a fast-moving body is also hindered by wave drag, as a swimmer creates surface waves like those produced by a moving ship.[124]

Why did we start swimming?

Our ancestors learned to swim perhaps as early as 100,000 years ago as they searched for food, but it was not until the 19th century that swimming became both a competitive sport and a widely practiced recreational activity, and the styles (freestyle, breaststroke, backstroke, butterfly) were formally defined. As with other competitive activities, there are major speed differences for swimming in open and enclosed water, and between short, medium, and long distances. The shortest indoor race is 50 meters (one Olympic pool length), the longest is 1,500 meters, and current male records for these distances (20.91 seconds and 14:30.67) translate to average speeds of, respectively, 2.4 and 1.72 m/s, no better than jogging. The speeds of elite swimmers at 10-kilometer open-water competitions have been stable, and the differences between males and females have been smaller than for other ultra-distance disciplines, with means averaging 1.45 m/s for men and 1.34 m/s for women.[125]

Harnessing our speed

There are three simple techniques for enhancing muscle-powered locomotion to reach speeds faster than running: ice skating, skiing, and cycling. The first skaters, more than three millennia ago, used strong horse or cow bones for sliding, and much later came medieval wood-and-iron skates strapped to shoes.[126] Those iron skates eventually acquired a frontal curl to prevent easy tripping: many winter paintings of the Old Dutch Masters are peopled with such skaters. By the 19th century the blades were all steel rather than iron, and they lengthened as they narrowed. The 20th century brought skating shoes with screwed-in blades, and leather was replaced by fiber glass, carbon fibers, and Kevlar: adult blades are now 46 centimeters long, hockey blades are 2.6–3.1 millimeters wide and speed-skating blades just 1–1.5 millimeters wide.

Experimental testing of reconstructed historical skate models has shown speeds increasing from about 2–4 m/s for the best medieval skates to 5–9 m/s for the early 20th-century designs. Modern high-speed stride frequencies are 0.45 per second and stride lengths can be as long as 20 meters. Comparing competitive skating and running records is easy, as the two disciplines use the same distances of 1,500, 5,000 and 10,000 meters. For the latter distance the skating speed record is more than 13 m/s, or almost exactly twice as fast as the best running time for that distance. Short-track skaters can reach speeds above 50 km/h (more than 14 m/s), the world record for the 500-meter distance is 53.56 km/h, and in 2022 Kjeld Nuis—on a 3-kilometer-long natural ice rink, skating within a "wind-catcher," a plastic enclosure to keep him out of the wind, attached to a car—exceeded 100 km/h, reaching 103 km/h (28.6 m/s).[127]

Cross-country skiing can never be as fast as skating or, of course, downhill skiing. After centuries of using the simplest wooden skis, primitive bindings, and ordinary footwear, modern cross-country skiing relies on well-designed skis, optimal-height poles, special bindings, purpose-built shoes, and (at the highest competitive level) elaborate training regimens to achieve unprecedented speeds.[128] Modern recreational cross-country skiers (in Norway, Sweden, and Finland they account for higher shares of the population than anywhere else) move at speeds of 10–16 km/h (2.8–4.4 m/s); competitive skiers, all of them now resorting to ski-skating rather than the classic Nordic style (a diagonal stride, lifting the back ski), move at average speeds of up to 25 km/h, significantly faster than runners over comparable distances.[129]

When Vasaloppet, the world's largest cross-country ski race (90 kilometers, with a maximum of 15,800 participants, run the first Sunday of March), was established in 1922, the winning times during its first decade were commonly longer than six hours or as slow as 4.1 m/s, but its 2022 record run was just 3:32.18, or 7.06 m/s.[130]

The fastest time for Norway's annual Holmenkollen 50-kilometer cross-country skiing race was posted by Sjur Røthe in 2015 at 1:54:44.9—an average of 7.2 m/s compared to 5.6 m/s for Kiptum's 2023 marathon record.[131]

In downhill skiing, competitors convert potential energy to fast motion on minimum-friction surfaces. There are now many kinds of such races (downhill, super G, slalom, Alpine combined, parallel) at speeds of up to 160 km/h, but the fastest on-ski speeds are reached by racers plunging straight down steep inclines on special skis (2.4 meters long) clad in air-tight polyester/spandex suits (with foam fairings on the lower legs), aerodynamic helmets, and using bent ski poles. Record speeds for this event are 254.958 km/h measured over a course of 100 meters.[132]

Wheels of change

The bicycle may be a simple machine—a sturdy tubular metal frame, handlebars, two equally sized wheels, a chain-driven back wheel— but this practical version is a remarkably late invention: it came only during the late 1880s, a century after Watt's improved steam engine, and was preceded by Edison's commercial generation of electricity.[133] Earlier bicycles were not just impractical but outright dangerous contrivances, none more so than the penny-farthing ("ordinary") bicycle of the 1870s, with its oversize front and miniature back wheels. The bicycle became popular only after John Kemp Starley and William Sutton introduced their Rover series in 1885 (equal wheels, diamond-shaped frame), and by 1889 they were improved by the adoption of inflatable pneumatic tires (instead of solid tires) and back-pedal brakes.

Intricate rear derailleurs came later, but fundamental changes began only during the late 1970s, with both very light designs made of expensive alloys and composite materials and, for speed racing, with solid wheels (for reduced friction) and upturned handlebars, as well as sturdier mountain bikes with fat tires, suspension forks, and disc brakes for riding not just off-road but also on some of the most forbidding narrow mountain paths. The speeds of mountain bike races are obviously limited by the prevailing terrain, and they do not usually average more than 15 km/h. Every other form of bicycle racing is considerably faster.

The top riders of popular (and chronically doping-infested) European races (Tour de France, Giro d'Italia, Vuelta a España) average about 40 km/h across the entire course, with climbs (often extremely punishing) at less than 20 km/h, the fastest (time trial) flat runs at 55 km/h, and brief spurts toward the finishing line as fast as 65–70 km/h.[134]

These speeds are routinely surpassed during track racing: wood-floor velodromes are steeply banked (maxima of 45 degrees) oval tracks where riders reach a maximum of around 87 km/h, and one-hour riding records are 56.792 kilometers for men and 49.254 kilometers for women. For comparison, elderly suburban cyclists taking their evening ride will move at less than 15 km/h, while younger riders will do 25 km/h. Listings of bicycle speed records also contain two highly contrived (and much faster) categories: motor-paced racing (inside a fairing pulled by a car) and downhill rides on snow. The record for the former (on a flat surface) is very close to 300 km/h, for the latter (on a serial production bicycle) about 210 km/h, but these are mere ephemeral curiosities, not genuine cycling speeds sustainable for longer periods of time.[135]

And it was the combination of cycling and advanced light plane design that eventually made it possible for humans to be airborne by using only their muscle power. Many (some tragically ended) trials proved that even the greatest possible exertion of highly trained adults does not suffice to become airborne by mimicking the flapping flight of birds. Humans are too heavy (the heaviest flying bird, the great bustard, weighs up to 20 kilograms), their surface area is too small in relation to their body mass, and they have no membranes (patagia) acting as airfoils to be able to glide as some small mammals (flying squirrels, colugos) can.[136] Unaided flying will thus always remain the only major type of locomotion that will elude human capabilities, and after the Second World War it became clear that the only way to become airborne even for short periods of time is by furious pedaling.

But even a top cyclist would not have succeeded unless engineers and aerodynamicists were able to come up with winged structures large and light enough to keep their mass and the rider's body weight aloft (even if only barely, flying just above the ground or above water) for long enough to meet the requirements for winning prizes for the first human-powered locomotion through air. The first man-powered flight was achieved only on August 23, 1977, when Bryan Allen pedaled fast enough to lift the *Gossamer Condor* off the ground, fly a figure eight for 1 mile, and clear the 10-foot hurdle at the end of the course.[137]

The large and light (96-foot-long and 70-lb) wing made of aluminum and a thin layer of Mylar was designed by Paul MacCready and Peter Lissaman, and Allen completed the course in 7 minutes and 27 seconds to win £50,000. That speed, 3.6 m/s, was the equivalent of moderately fast running (3.7–4 m/s). Later, longer distances were flown by pedaling, including the crossing of the British Channel by Bryan Allen in the *Gossamer Albatross* in 1979 (at an average speed of 3.5 m/s), and in 1988 by Kanellos Kanellopoulos from Crete to Santorini (115.1 kilometers) in *Daedalus* (made from carbon fiber tubes, polystyrene, and Kevlar), which averaged 6.9 m/s (24.8 km/h, faster than the fastest marathon run) before it crash-landed into the sea less than 10 meters from the shore.[138] Other attempts to break strange bicycling records, both airborne and on the ground, followed. In 2015, Robert Förstemann, a track cyclist, pedaled furiously to sustain a power of 700 watts just long enough to slightly brown a slice of bread in a toaster, and in 2019 Guy Martin failed to cross the English Channel by pedaling a bike suspended from a helium-filled balloon as big as a bus. What next?

3. Speed Limits in Premodern Societies

All prehistoric foraging activities—as well as foraging in sedentary agricultural societies, practiced to enrich staple diets with wild plants, mushrooms, and wild meat—proceeded at speeds that were inherently limited by natural constraints on mammalian metabolism and on the efficiency and endurance of human and animal muscles. Hunting and gathering was restricted by speeds of walking, running, and paddling. Walking dominated the daily activities of all foraging societies, not just during searches for edible plants and sedentary or slow-moving organisms (mollusks, turtles, insects, and their larvae), but also during repeated moves of entire groups that were undertaken periodically in search of better foraging opportunities—often as many as a dozen times a year, and sometimes across considerable distances.[1]

Besides brief runs occasioned by many daily activities, in the previous chapter we saw the convincing evidence of endurance running in pursuit of prey. Speed was of secondary importance in this effort, as no hunter could hope to outrun a large ungulate trying to escape its pursuers. But the human ability to thermoregulate by sweating during hot weather made it possible to run the animals down, often after covering distances of many kilometers (with the pursuit also requiring expert tracking), due to the overheating of the prey's bodies and ensuing prostration. Endurance and persistence defeated what seems, initially, the insuperable speed advantage of the swiftest antelopes. In contrast, paddling—even with some remarkably efficient designs of kayaks and canoes—is no faster than a fast walk, as it must deal (even when water is calm and when there is no contrary wind) with water's considerable resistance.[2]

Rudimentary food processing and storage (cutting and grinding roots; hoarding nuts; butchering animals and scraping their skins; drying or smoking meat and fish) could proceed only as fast as allowed by the repetitive muscle exertions of skilled individuals.[3] The most fundamental consequence of these realities was the limited number of people that could be supported by these activities. The enormous range of foraging environments (from tropical rainforests to treeless Arctic coastlines) makes it impossible to quote any typical durations of daily foraging activities, but the notion that these small groups (typically 20–40 people) constituted "the original affluent society"—or as put forth in many recent popular books, worked less, ate better, and were happier than us today—is an excessive generalization relying on limited and selective findings rather than an impartial conclusion based on all available evidence.[4]

Those familiar with the concept's history might conclude that it was primarily a deliberate ideological construct rather than a universal characterization that could withstand closer inquiry into the actual lives of foragers in a wide range of environments.[5] The best available evidence indicates foraging densities spanning three orders of magnitude, with minima as low as one person per 100 km^2 and with typical rates almost always well below one forager per square kilometer: the latter density would be like having fewer than 20 people roaming Manhattan or fewer than 300 foragers surviving in the area occupied by Tokyo's 23 wards (619 square kilometers, now with 9.4 million people).[6]

The gradual adoption of a sedentary existence tied to the regular cultivation of annual or permanent crops and the domestication of (eventually) a dozen mammalian and avian species allowed for much higher population densities and for the creation of food and material surpluses that could be channeled to activities other than subsistence food production. To begin with, these surpluses were small but in suitable environments (including Egypt, with its guaranteed annual Nile inundation bringing a reliable supply of moisture for annual crops) they allowed for the creation of the first stratified societies, complex belief systems, formalized rituals, and monumental architecture as early as nearly 5,000 years ago. Egypt and other ancient civilizations, including those of Mesopotamia, India, and the northern Chinese states, became the first complex societies whose tempo of life was determined by the normal speeds of humans and quadrupeds—that is, ultimately by the speed with which those societies could produce their staples.

Grains were the most important as well as the most reliable source of food, while large, domesticated mammals harnessed for work sped up the production of food as well as the transportation of heavier loads on land. And both people and animals were liberated from engaging in some repetitive, labor-intensive tasks by the introduction and gradual improvement of the first two inanimate energy conversions, turning the kinetic energies of water and air into useful work and speeding up tasks ranging from milling grain to raising water, and from sawing wood to hammering iron. Initially these were not stunning gains, but eventually they became large enough to allow slowly rising shares of the population to engage in activities other than food production, and to provide the foundations of modernity.

This was, even more so, the case with the gradual advances in long-distance sailing. For millennia, the designs of sails and ships and navigational practices limited these capabilities to disconnected or only marginally interacting regions: the Roman Mediterranean (*mare nostrum*); commerce (later Muslim-controlled) in the Indian Ocean; trade in the East and South China seas. Then, again, a combination of technical improvements eventually produced ships able to sail to distant continents, circumnavigate the world, and start setting up the first limited but indisputably globally oriented economies (Spanish, Portuguese, English, Dutch), whose development reached a considerable extent before the age of mechanization.

Concurrently, premodern mechanization based on water and wind power (waterwheels and windmills of different designs and capacities) sped up many previously highly labor-intensive manufacturing tasks, leaving room for further specialization and product differentiation and liberating human labor for other manual activities (construction, transportation) or for tasks where intellectual rather than physical capabilities were paramount. During the early modern era these much-expanded activities ranged from efforts to improve existing machine designs and to invent new modes of production to book publishing on unprecedented scales, an activity that opened the way for an acceleration of scientific and industrial advances.

Elaborate water pumping works built to deliver water from the Seine to the Palace of Versailles, the largest project of its kind during the 18th century, were an outstanding example in the category of traditional (still overwhelmingly wood-based) mechanization. The *machine de Marly* had 14 large waterwheels (each with a diameter of nearly 12 meters) whose rotation powered more than 250 pumps, to raise water by 162 meters to supply the Versailles aqueduct.[7]

The growing speed of book publishing was due to a much faster rate of typesetting that was made possible by movable type cast in an alloy of lead, tin, and antimony. European printers brought out 11,000 new editions between 1455 (when Gutenberg began printing his Bible with movable type) and 1500; a century later, the 50-year rate rose more than 13-fold, and during the second half of the 18th century it reached nearly 650,000 new titles.

And while 35 million copies of books were published in France between 1525 and 1575, two centuries later the total had more than quadrupled to in excess of 157 million books.[8]

Grains, quadrupeds, and wooden machines: the world of slow speeds

First things first: this review of speeds in premodern (before 1500) and early modern (1500–1800) societies will start with the fundamentals of survival, the production of domesticated crops, and the management of animals kept for milk, meat, eggs, and draft. The most obvious fundamental limit was the time needed by plants to grow from seeds (or tubers or divided roots or cuttings) to maturity, and for animals to reach the age at which they could be milked or slaughtered for meat. Domesticated crops are, overwhelmingly, annuals, and cereals—complemented by legumes, tubers, and oil crop—supplied most of the desired combination of carbohydrates, proteins, and oils needed for the balanced diets of the Old World's high civilizations, as well as in Mesoamerica and in Andean societies (and also for the feeding of non-ruminant domestic animals).[9]

Staple cereals needed, without exception, at least four months (often five) to mature for the spring-planted crops (for example, European spring wheat, planted in April and harvested in August) and 10–11 months for overwintering crops planted in late summer or early fall (such as German wheat and rye, planted in September or October and harvested between June and August).[10] In China, winter wheat was planted in September and October and harvested in June; single-crop rice went in during April and May and was harvested in September and October.[11] This inherently slow pace of staple crop maturation contrasted with the relatively rapid production of vegetables—less than two months for lettuce; less than three months for beans, cabbage, peppers, or tomatoes—but they (with a water content in excess of 90 percent and with a mere trace of protein and no fat) could supply only vitamins (mainly C and A) and minerals, not macronutrients.

Pastoralist societies could, in times of food shortages, kill more animals than they would do in normal years, but they had to be careful not to reduce the long-term reproducibility of their herds. For sedentary cultivators, that was a very limited option for at least three reasons. First, in smallholdings numbers of large ungulates (donkeys, oxen, horses) were always limited. Second, the long periods required for gestation, maturity, and training (to become reliably docile) meant that it would take several years to rebuild the desired number of working animals by using the remaining breeding stock. Third, unlike today when pigs go from weaning to slaughter in less than six months and chickens in less than six weeks (kept in confinement and fed optimal diets), traditional pigs and poultry were free-roaming animals, fed waste scraps, and herded to graze in forests and fields, and it took well over a year for pigs and three or four months for chickens to be slaughtered—periods too long to provide any rapid, repeated food relief.[12]

Although modern plant breeding has produced several early maturing crop varieties, and although the combination of breeding, confinement, and optimized feeding has substantially cut the time to maturity for poultry and pigs, the fundamental life-cycle constraints of plants and animals still matter in all modern societies. But (as I will explain in the next chapter) we now have effective means (long-term large-scale food storage and refrigeration; global trade in all staples) to minimize or negate the effects of much-reduced or failed harvests, as people in the stricken areas do not have to wait for months until a new harvest delivers a better food supply. As a result, large-scale famines have become history, and undernutrition and hunger would be much reduced without recurrent civil wars that impede food aid deliveries.[13]

But because of the absence of any substantial interregional trade and the lack of long-term grain storage in nearly all premodern societies, every below-average harvest meant undernutrition and hunger, and failed harvests brought recurrent famine. Rome, and other populous coastal cities of the Roman Empire, relied on substantial exports of wheat from Egypt, Libya, Tunisia, and Sicily (as well as on a similarly extensive trade in olive oil and chickpeas), but this trade ceased with the empire's dissolution and, soon afterwards, the Muslim conquest of the entire southern Mediterranean littoral, and the demand for feed crops (alfalfa) for horses also declined as the empire's armies and its famous courier and transportation service (*cursus publicus*) ceased to function.[14]

Reap what you sow

A notable adjustment resulting from these changes was a more common cultivation of rye in early medieval Europe: its seeds could be sown in fall, after the wheat was brought in, and the crop was harvested earlier than wheat, speeding up the timing of a new staple harvest by 1–2 months. And while it was not a near equivalent of preferred bread wheat, rye could be leavened with sourdough to produce a heavier but nutritious bread.[15] In the absence of a life-saving large-scale interregional grain trade, the best way to cope with recurrent crop failures—and hence with the inherently long wait for the next harvest—was to set up sufficient stores. China was the only major traditional civilization that developed this requirement as a matter of continued state policy.[16]

In any case, the only way to return staple food supply to an expected level—that is, to assure a better new harvest—was to start with plowing to prepare the ground for planting.[17] Plowing was an indispensable procedure, but could proceed at only limited speeds and required much physical exertion by both men and animals. Many natural soils are heavy and full of dense (wild plant) roots, as are many repeatedly cultivated soils (which are full of weed roots), and horizontal clods of soil must be displaced before planting. That could be done, laboriously, by hoeing. Traditional hoe designs differed in the blade shape (commonly rectangular, but also triangular or circular), handle and blade lengths, and blade width.

Hoeing speed depends on the tool's weight, the angle at which it strikes the soil, the height to which it is lifted, and soil density. Assuming a work-to-rest ratio of 10:7, low-lift hoeing can prepare a bit more than a square meter of field per minute; high-lift hoeing in heavier soils will accomplish no more than about 0.33 m²/min.[18] The latter rate would require about 500 hours, or 50 10-hour days of exertion per hectare, a speed that is obviously too low to deploy the technique beyond cultivating small patches of land, typically for vegetable crops. Large areas could be cultivated only by plowing, and that method of land preparation cannot be energized by human exertions. Even shallow plowing in lighter soils requires a steady pull of 80–120 kilograms, and as prolonged draft power typically amounts to 15 percent of body weight (although for short periods it can be up to 25–35 percent), such a task is well within the capabilities of two 400-kilogram oxen or a single large horse, but is beyond normal human means: even four adult men could not sustain such effort.[19]

While gardens can be hoed, field-size crop cultivation became possible only with animals harnessed to plows. The first plows were just simple shaped pieces of wood capable of little more than scratching the soil or opening a small furrow. Only the adoption of moldboards— initially just angled pieces of wood that pushed the plowed soil to one side, turned it over, and buried the cut weeds—led to effective plowing. Heavy wooden moldboards were eventually substituted by curved iron (and after 1860, steel) plowshares.[20] Harrowing followed plowing, to break up larger clods and to smooth the soil surface for seed planting.

Quadrupeds harnessed to plows in different parts of the world have included camels (with an inefficiently high point of traction), yaks, cows, and donkeys (the latter two being both too slow and too weak)—but oxen (castrated adult bovine males) and water buffalo have always been the dominant draft animals in subtropical and tropical regions, while in temperate latitudes oxen were eventually largely displaced by horses (and in the US also by mules, which are donkey-and-horse hybrids). Oxen were less powerful than horses, but easier to feed and less expensive to harness (by simple wooden yokes). But they were always slow, and using them for plowing and harrowing was nearly always the most time-consuming task in traditional crop cultivation.

Horse power

A simple Roman scratch plow (*aratrum*—originally wheel-less, later with two small wheels) had a symmetrical share and created a shallow furrow without inverting the soil, making it necessary to cross-plow the field at right angles to the first pass.[21] Assuming a furrow width of 20 centimeters and a walking speed of 0.8 m/s, a hectare of land (100 × 100 meters) would have required about 17 hours to plow in a single direction (that is, nearly 10 m²/min, 30 times faster than hoeing), and the farmer and his two oxen would have to walk some 50 kilometers. Draft animals have always varied in size and performance, and eventually a single heavy draft horse could easily surpass a pair of good oxen.

Even though the gait of horses working in fields (or in heavy transport and in mines and factories) was limited to a walk, often to a slow and laborious one, the animals were almost always significantly faster than oxen. Their typical working field speeds were around 1 m/s (mostly between 0.9 and 1.1 m/s, with the range valid also for mules) compared to 0.6–0.8 m/s for oxen and water buffalo and 0.7 m/s for donkeys, but a recent study put the speed of working

Ethiopian oxen at just 0.4–0.5 m/s.[22] The most important consideration was the greater draft power of horses—but not of all of them. Body weight comparisons have relatively large, and sometimes overlapping, ranges.

Quoting a truly representative average weight of the premodern horse is impossible, but it is safe to conclude that most adult animals could not sustain work at the rate of 1 horsepower (about 745 watts). A range of 500–600 watts would include most of the power ratings for horses used for field work before the 19th century, compared to about 300–400 watts for most oxen and 250 watts for donkeys, but turning horses into efficient draft animals required three preconditions: efficient harnesses, horseshoes, and proper nutrition. A horse's considerable power could not be deployed efficiently without a comfortable harness to transfer the animal's exertion to the task of plowing, pulling a wagon, or raising water with the maximum possible efficiency.[23]

The dorsal yoke, placed directly behind the withers and held in place by straps, was used to harness horses in antiquity. This was followed by a breast collar that originated in China in the 3rd century BCE—another Chinese invention that eventually reached Europe and every continent. The main component of the collar harness is a single oval wooden—later metal—frame ("hame" is the old English term) that is lined or padded to fit comfortably on the horse's shoulders, and draft traces (the straps which run from the collar to the load) are connected to the hame just above the horse's shoulder blades.

Horse hooves evolved on grasslands, and they were not optimized for working in heavy wet soils: they would swell and split. The dating of the horseshoe's origins remains uncertain: what is clear is that in Europe shoes were in common use by the end of the first millennium CE and mass-produced a few centuries later. As with the development of almost all practices involving animals, we have to think about the scale of trial and error needed to perfect hammering horseshoes into the hooves of a large and powerful horse.

Unlike cattle, horses are not ruminants (with a multi-compartmented stomach able to digest large amounts of forage) but rather non-ruminant herbivores. Parts of their large intestine are like a cow's rumen and hence they can consume and digest forage rather efficiently, but for the best performance their diet should be complemented by grain, with oats having the best nutritional profile for working horses. In the early 20th century, when the numbers of working horses in the US reached record totals of more than 20 million animals, the best feeding recommendation was a daily ration of 4.5 kilograms of oats and 4.5 kilograms of hay.[24]

But even well-fed, collar-harnessed, and shod premodern draft horses were not equal to their 18th- and 19th-century descendants: not as well fed, they were generally smaller (shorter by 3–4 hands), and hence less powerful and unsuited for the plowing of heavy clay soils; moreover, rudimentary veterinary medicine could not save many of them from premature death. The transition from slower and less powerful (but able to work longer hours) oxen to horses was gradual: in England horses came to dominate field work only by the late 1500s; in poorer parts of Europe oxen remained important even during the 19th century.

Marginally faster: a collared horse pulling harrows (and seeding by broadcasting, with birds eager to eat the seeds) from the early 15th century's *Très Riches Heures du Duc de Berry*.

Plowing through time

Variance in the type of soil, its moisture content, the kind of plow, the width and depth of cultivation, and the age and health of animals make it impossible to quote a typical plowing speed, but the most common range for a late-19th-century two-horse team and a man guiding a steel moldboard walking plow was 0.9–1.3 m/s or about 3.2–4.7 km/h—up to roughly three times faster than using a pair of oxen, whose typical plowing speeds with lighter moldboards were no more than 1.8 km/h (0.5 m/s). Assuming an average working speed of 1.1 m/s and a 30-centimeter-wide furrow, a farmer and his two horses would have to walk about 33 kilometers in 10 hours to complete the task, averaging nearly 17 m²/min—more than double the speed of oxen-powered plowing.

Draft animals were also indispensable for harrowing, for bringing harvested grains (commonly bound in sheaves) to threshing grounds, and in many cases for threshing, with oxen driven in circles over spread grain, sometimes pulling a wooden board, separating ears of grain from stalks and grains from husks. Sowing and harvesting were done manually—by broadcasting seed and using sickles or scythes, with some crops cut below the ears and others near the ground (requiring tiresome bending and stooping)—with both seed drills and mechanical reapers deployed widely even in Europe and North America only after 1850.[25]

We have enough historical information to reconstruct typical labor budgets for the production of wheat, the staple of European and Middle Eastern food supply and the dominant grain of the northern regions of India and China, and hence to calculate the sequence of wheat production speeds across the span of two millennia, from Rome's early imperial period (the 2nd century CE) to America's most productive wheatlands of the 2020s.[26] I have reduced my sequential calculations to a single, easily comparable quantity—the rate of producing wheat per hour—or to its reciprocal value, the time needed to produce a kilogram of wheat.

In Roman Italy, a farmer preparing a field with two oxen, a simple scratch plow, and wooden harrows, harvesting with sickles and threshing the crop with the help of oxen, spent about 180 hours to produce about 400 kilograms of grain from a hectare (four *iugera*) of wheat. That works out to about 2.2 kilograms per hour, or roughly 25 minutes per kilogram of grain. In 1200, a millennium later, an English farmer working again with two oxen, but with better implements, had to plow in heavier soils, threshed the grain by flailing, and harvested at least 20 percent more than his Roman predecessor. He produced about 3.2 kg/h, and required some 18 minutes per kilogram of grain. Half a millennium later, a Dutch or English farmer—using two good horses, a moldboard plow with iron share, and a seeding drill and scythes for harvesting—harvested (even after spending considerable time fertilizing his field with manure) 2 tons of grain per hectare: that translates to 12 kg/h or to just 5 minutes per kilogram of wheat. During the 19th century, wheat farming everywhere remained powered by human and animal labor, but the latter became dominant as greater exertion was needed to pull heavier machinery. By 1900, American farmers had opened large areas of new farmland on the Great Plains only thanks to a new, hybrid form of field work. Animals still supplied all draft power, but the implements they pulled were no longer wooden—they were made of iron and steel that had become affordable thanks to rapidly unfolding industrialization.

Unlike his ancestors, a Great Plains farmer did not walk behind a heavy wooden plow; he was riding on a metal seat mounted above a wider steel moldboard that opened furrows up to 40 centimeters wide and was pulled by four strong, well-fed horses. Other steel implements followed the plowing to prepare and plant a field (harrows and a wide seeding drill), and the crop was harvested with a mechanical reaper-and-binder and threshed with a horse-powered threshing machine. The entire sequence took just 22 hours of labor, and with an average yield of just 1 t/ha in the US in the 19th century, this translated to 40 kilograms of grain per hour of labor, or merely 1.5 minutes per kilogram.[27]

These calculations indicate that at the beginning of the 20th century the combination of human and animal exertions could produce a unit mass of staple grain about 17 times faster than two millennia ago—or, conversely, that the entire sequence needed about 94 per-cent less time to accomplish! It would be reasonable to question how accurate, or how representative, are the reconstructions of ancient Roman labor expenditures I used to calculate the wheat production speed of some two millennia ago. But remarkably, we have excellent confirmation of this value thanks to a unique collection that illus-trates these nearly immutable commitments both accurately and in detail.

Between the two world wars, John Lossing Buck, an American agricultural economist working in republican China, organized—and between 1929 and 1933 headed—by far the most detailed study of traditional farming ever. It gathered data (ranging from land own-ership and dietary intakes to field labor and wages) primarily from personal in-depth interviews and on-site observations conducted in 22 provinces and nearly 17,000 farms across the country, from Gansu in the arid northwest to Guangdong in the subtropical south. The results, published in Shanghai in 1937 (with data in large folio format), offered detailed quantitative perspectives of every aspect of China's traditional farming, which at that time was still based solely on human and animal labor—typically with just a pair of *huang niu* (yellow cattle) or *shui niu* (water buffalo) used for plowing and harrowing—just as it had been during the centuries of previous dyn-astic rule, and also in Roman Italy two millennia ago.[28]

The average time to produce wheat on a hectare of land in China's main wheat-growing region was (in terms of adult male labor) about 400 hours, and the yield (thanks to the recycling of animal manure and human waste) was close to 1 t/ha, a performance that translates into producing 2.4 kg/h, or labor needs of 24 minutes per kilogram of wheat, just slightly better than the reconstructed Roman mean. These reliable early 1930s Chinese accounts of labor in crop cultiva-tion provide convincing demonstrations of the inescapable limits on reducing labor needs in systems constrained by human and animal capabilities.

And these accurate Chinese accounts can also be used to show how the substantial speed advantage that resulted from having teams of heavier horses harnessed to good plows, harrows, seeders, and harvesters could be largely, or completely, negated by the larger size of cultivated fields. A Roman farmer who plowed 20 *iugera* of land (5 hectares, a holding well above the minimum subsistence requirements) with a team of his oxen and a 15-centimeter-wide wooden plow that did not cut too deep, had to walk some 330 kilometers behind his plow and, with an average speed of 1.8 km/h, needed about 180 hours to finish that task.

In late-19th-century America, a Great Plains farmer used powerful horses and better plows but, in absolute terms, the size of his holdings made this basic cropping operation no less time-demanding. Plowing 40 acres (about 16 hectares) with a team of two horses attached to a steel moldboard plow making a furrow 12 inches (30 centimeters) wide would have required the farmer to walk about 530 kilometers. In heavier soils, at 3.2 km/h, that would have taken about 165 hours. Obviously, the plowing speed had nearly doubled, and the harvest (everything else being equal) had more than tripled, but the total time spent on that essential farming task had barely changed—and the comparison would be even less favorable if we considered the fact that a 19th-century farmer had to devote a significant share of his plowed land to growing feed for his horses.

The extent of this diversion (feed instead of food) is best illustrated by reliable American data for the first two decades of the 20th century, when the number of American draft horses (and mules) peaked at about 25 million. Growing enough feed (oats, corn, and hay) for their adequate nutrition (including the feeding of non-working animals required for breeding and herd maintenance) required close to a quarter of America's large expanse of cultivated land.[29] In addition to these considerable feeding costs, there were other constraints on speeding up field tasks by deploying more animal power: very high peak-power requirements, and the logistics of harnessing and guiding increasing numbers of animals to perform previously impossible tasks.

The first combine (harvester and thresher), patented in the US by Hiram Moore in 1836, cut a swath of 4.5 meters wide and had to be pulled by up to 30 horses or mules, with a ground-driven bull wheel powering the thresher's moving parts.[30] Later (during the 1880s) came another hybrid design: a horse-pulled combine whose moving parts were powered by a steam engine, which eventually cut wheat 15 meters wide and required up to 40 strong animals in harness.

That was equal to 30 kilowatts of draft power, and a hectare of wheat could be cut and threshed in only about 10 percent of the time needed by the traditional laborious sequence of cutting (by a two-horse team with a reaper), binding, shocking, moving sheaves, and threshing—but the number of required animals and complications of their harnessing detracted from the advantage of a speedier harvest. Only tractor-powered plowing (and harrowing) and harvesting by combines (first tractor-drawn, later self-propelled) resulted in absolute speed-related time gains, and also eliminated the need to devote cropland to growing animal feed and to assemble unwieldy draft teams.

In contrast to North America and most of Atlantic Europe—where faster, more powerful, better-fed, well-shod, and better-harnessed horses displaced oxen in field work during the 19th century—most of Asia's cropping operated under centuries-old constraints even during the early decades of the 20th century. Basic tasks were not sped up, as the time required for their completion was essentially fixed because productive capabilities remained stagnant. Draft animals were used only for plowing, harrowing, and threshing, while other activities—including the transplanting of rice into wet fields, crop weeding, the highly labor-intensive fermenting of manure and human waste and carrying them to fields in buckets hanging on shoulder poles, harvesting, and taking grain sheaves to threshing grounds—continued to depend on human exertions.

But some traditional societies had made important advances beyond the waiting spells dictated by the maturation of crops (mostly 3–5 months) and beyond the speeds and power limited by human and animal metabolism (speeds mostly at slow-walking pace; power limited usually to the equivalent of a few working animals). Waterwheels—machines that became common during the early centuries of the common era—were the first inanimate energy converters that harnessed the kinetic (and potential) energy of water for performing many tasks that had previously required tedious and

Heavy wooden post windmills were ubiquitous in early modern Holland:
this one was painted by Jacob van Ruisdael.

exhausting human and animal labor: their most common use was for
grain milling, but eventually they were adapted to tasks ranging from
wood and stone sawing to fulling cloth, hammering metal in forges,
and lifting water from wells and low-lying fields.[31] A few centuries
later (during the Middle Ages) windmills became the second type of
widely used and relatively powerful inanimate energy converters.[32]

These two fundamental innovations succeeded in both speeding
up many repetitive tasks and multiplying the power available for
their execution. But while their contributions should not be underestimated, their adoption was uneven: only in some regions did the
concentrations and capacities of these simple wooden devices reach
levels that resulted in notable gains in food output or artisanal production. The two conversions also differed in some fundamental
ways, first because of the different nature of the two media (water
and wind), and then because of mechanical limits in designing and
building wooden machines capable of converting those natural kinetic energies to useful work.

Early designs of waterwheels included both horizontal machines
and vertical devices: the former had a common shaft directly rotating
the millstones positioned above them. The most remarkable were the
medieval horizontal wheels in Spain and Persia that worked under

heads (the vertical distance between upper and lower water levels) of up to 8 meters: with water guided through a wooden jet (with a variable opening) those wheels could rotate 150–160 times per minute, and could grind up to 150 kilograms of flour per hour.[33] Vertical machines (with horizontal axes) could be undershot wheels (easiest to build, with water flow impacting their submerged paddles), overshot wheels (more expensive to build, requiring conduits to deliver water from above the wheel to the buckets), or an in-between design (breast wheels) with water impacting the wheel at the 10 o'clock position.

Obviously, all vertical machines required gearing to transform the vertical motion to the horizontal grinding of grain by millstones and to accelerate their speed: overshot wheels rotated slowly, mostly 2–10 turns per minute, while grindstones had an optimum speed of 175–200 revolutions per minute. With all gears being wooden, and wearing out under heavy use, the resulting friction detracted from the overall efficiency. The high density of water ($1,000$ kg/m^3) means that even streams with moderate speeds have considerable generation potential due to their substantial kinetic energy, and the construction of overshot wheels (often fed by long canals) enabled good performances even with slow water flow.

But for hundreds of years after waterwheels became important components of premodern economies, they were overwhelmingly wooden devices—with all that implies for efficiencies of their everyday operation. Theoretical calculations put the maximum efficiencies of overshot wheels at 71 percent and undershot wheels at 30 percent, while actual 18th-century measurements (based on what were considerably more advanced designs compared to the machines of late antiquity) showed performances of, respectively, 63 percent and 22 percent.[34] In the absence of any actual reliable power ratings, the best way to quantify the water-powered acceleration of common tasks is to turn to information on the typical hourly outputs of grain-milling practices.

The tiresome hand-grinding of wheat with quernstones, mostly done by women, would produce 2–3 kilograms of coarse flour per hour; during the late Roman republic (before waterwheels became common), slaves turning *mola manualis*, an hourglass-shaped stone mill, would produce no more than 7 kg/h, and larger *mola asinalis*, turned by a harnessed donkey and employed by large Roman bakeries, would produce 10–25 kg/h—but even a small water mill (whose power was equal to that of two strong horses, or about 1,500 watts) could grind about 80 kilograms of flour in one hour, and in late imperial Rome there were many larger mills that could reliably produce 150–200 kg/h.[35] During the Middle Ages, undershot waterwheels were installed on boats anchored in rivers flowing through some European and Middle Eastern cities. Typical sizes gradually increased, and reliable performances improved, and many water-powered mills could produce in one day enough flour to feed 2,500–3,000 people.

And we have a detailed understanding of what was, undoubtedly, the technical pinnacle of water-powered milling, designed in 1785 by Oliver Evans, an American inventor who began his career as a wheelwright and wagon maker and ended it as the inventor of high-pressure steam engines. His was the first fully automated flour mill, which used water power not only to grind grain but also to handle its transfers by hopper-boy, bucket elevators, conveyor belts, and Archimedean screws. These machines performed "every necessary movement of the grain, and meal, from one part of the mill to another, and from one machine to another, through all the various operations, from the time the grain is emptied from the wagoner's bag . . . until completely manufactured into flour . . . without the aid of manual labor, excepting to set the different machines in motion."[36] His design received the third-ever US patent and was described in *The Young Mill-Wright & Miller's Guide*, a self-published book that came out in 1795 and saw 15 editions by 1860.

In 1791, George Washington, at that time serving his first term as the president, upgraded the mill at his Mount Vernon plantation by installing all mechanical improvements designed by Evans. Washington's mill was powered by a pitchback wheel (rotating counterclockwise), with water from a flume impacting the wheel at the 10 o'clock position.[37] Only one man was needed to run the operation, but six men were employed to close the flour-filled barrels: the mill was able to grind about 7.5 tons of wheat a day, producing more than 6 tons of flour or about 600 kg/h in a 10-hour day.

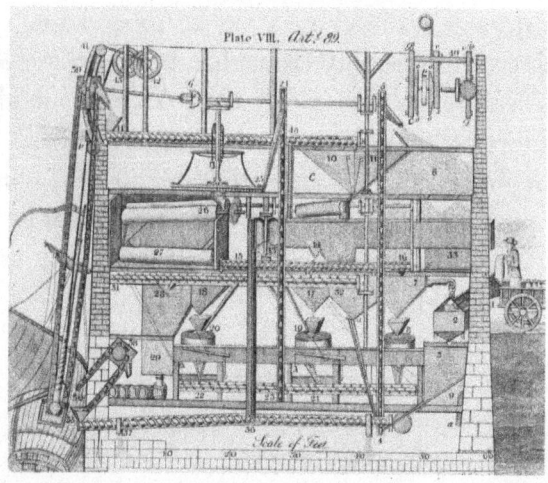

Plate VIII.

Scale of Feet

The fastest premodern way of milling wheat: a complete depiction of Oliver Evans's ingenious water-powered design.

In early imperial Rome, it would have taken nearly 100 slaves turning *mola manualis* to match that hourly output—but that is only if we make a simplistic comparison by dividing the two production rates. Milling done by slaves was even more burdensome, because a complete analysis would have to consider the labor needed to produce food for those hard-working men, and because an even larger labor force would be needed as it could not be expected that, unlike a well-built waterwheel, those men could work 10 hours a day, day after day, without getting sick or injured. That is even more obvious given the near-hellish conditions of their work: poorly clothed, branded on foreheads, with leg irons, backs scarred from lashing, and skin and eyes sore due to oven smoke and flour dust.[38] This is a point to which we could return again and again as we trace the historic gains in speed: quite often these improvements went far beyond accomplishing specific tasks much faster; they also truly changed lives thanks to their attendant qualitative improvements.

And so did the exploitation of wind, the other kind of inanimate kinetic energy whose conversion to useful power was mastered by many traditional societies when they built and operated a variety of windmills. Air is, of course, a very different medium to exploit than water: it is nearly two orders of magnitude lighter (the respective specific densities at sea level are 1.225 kg/m³ and 1,000 kg/m³), but wind speeds are considerably faster than typical water flows. While moderately fast streams have speeds of 1.2–1.5 m/s, a moderate breeze (recall Beaufort's wind scale) is nearly 7 m/s and a fresh breeze has a speed of just above 10 m/s. That matters, because the wind energy of a parcel of moving air is the product of the area swept by a windmill's vanes (now by wind turbine blades), air density, time, and wind velocity raised to the third power:

$$\text{energy}_{\text{wind}} = 0.5 \cdot A \cdot \rho \cdot v^3 \cdot t$$

Consequently, even small increases in wind speed make a notable difference, but there are natural and design limits to the share of wind energy that can be converted to useful work (or, in modern installations, to electricity). In 1919, Albert Betz showed that no more than 59.3 percent of wind energy can be extracted by rotating machines, and in practice the efficiencies for early, heavy wooden windmills were no more than 30 percent.[39]

As with waterwheels, it took centuries before windmills reached larger capacities and were suitable for some high-performance tasks. While the diameter of small medieval windmill fans was no more than 10 meters, by the early 17th century the large octagonal wooden windmills of North Holland design had diameters up to 30 meters, the limit imposed by the length of available tree trunks.[40] These windmills were used for tasks ranging from grain and oil seed milling to wood sawing, but their most notable deployment was the draining of polders—low-lying tracts of land often created by the cutting of peat. The windmills not only sped up the necessary water-lifting, but as the following example illustrates, they enabled it to proceed on previously inconceivable scales.

In 1631, the town of Alkmaar was given permission to drain the Schermer polder with a total area of nearly 65 square kilometers. Jan Leeghwater, the era's leading hydraulic engineer, subdivided it into 14 plots and used 14 windmills to lift water into a storage basin, and from there 36 additional windmills (12 sets deployed in three stages) raised it into the ring-canal, lifting water at a rate of 1,000 m³/min;

the task was done between 1633 and 1635 and the reclaimed land has been farmed ever since.[41]

The largest 18th-century windmills working with brisk winds had a power of more than 20 kilowatts (or 25 horsepower), and many smaller mills rotating in average wind speeds could deliver 10–15 kilowatts. Depending on the wind's persistence, even a small mill could produce as much as 1–1.5 tons of flour daily—easily 100–200 times more than exhausting and tedious hand-grinding.

The introduction of metal gears during the 18th century made windmills more efficient, and the total number of mills kept on increasing during the 19th century as Europe's industrialization was sped up by the mass deployment of much-improved designs. Their numbers reached their peaks in Atlantic Europe just as the new internal combustion engines were becoming more efficient and more affordable. By 1900 there were about 30,000 windmills in the North Sea countries, most of them in Germany and the Netherlands, and the US had an even higher number of smaller windmills installed across the Great Plains.[42] But these peaks were followed by rapid retreats as combustion engines and electric motors took over—and wind-powered machines became popular once more only a century later, with the ascent of new large wind turbines.

On land and under sail

In the early 21st century, long-distance pedestrian travel is limited to extreme hikers and to some religious pilgrimages. In contrast, even in the richest parts of the world, walking was the dominant way of long-distance travel on land until the mid-19th century, no matter if the destinations were nearby (between two small towns, or from a village to the seat of a regular county fair) or if they took weeks, months, or even years to reach. Young adults could make some exceptional speeds over shorter distances, but the averages on longer trips all around the world had to fit within a narrow range of slow-walking speeds (about 4 km/h) that could be sustained for 6–10 hours a day and for many consecutive days. Extended European and Asian pilgrimages offer the best examples of this walking endurance.

Japan's most popular pilgrimage, around the island of Shikoku, began in the 12th century and settled into its final version of visiting 88 temples (*Shikoku Hachijūhachikasho Meguri*) by the late 16th century. Its complete version is about 1,400 kilometers long, and it takes, on average, 45 (35–56) days at about 30 km/day.[43]

Fewer *henro* (pilgrims) walk it now, as buses and cars have taken over, but hundreds of thousands of people annually still walk what is perhaps the most famous of all European pilgrimage routes, the Camino Francés: it takes the devotees from Saint-Jean-Pied-de-Port, a small French Pyrenean village, or from Roncesvalles (near the French border), via Burgos, León and Ponferrada to Santiago de Compostela in Spanish Galicia, a length of 790 (or 750) kilometers that is usually done by walking 6–8 hours every day for 30–34 days— that is, 23–26 km/day, a slow walk of no more than 4 km/h.[44]

Setting aside all long-distance prehistoric migrations (starting with modern humans moving out of Africa by about 90,000 years ago, and ending with the peopling of North America by more than 15,000 years ago), many prolonged walks during the historic era took travelers to other continents—individually as adventurers or pilgrims, and in large groups during mass-scale migrations (such as those of the Goths crossing the Danube to the Roman Empire after 376 CE and migrations of Slavic-speaking groups westward during the 6th century CE) or during military campaigns and wars of territorial conquest, waged with high frequency during the entirety of recorded history.[45]

Here is a single dated and averaged example from the Middle Ages, based on reliable information and confirming the expected slow pace of such peregrinations. Starting in the late 11th century, hundreds of thousands of European crusaders walked across most of the continent to Turkey and then on to Syria to reach the Holy Land. While knights rode their small medieval horses (and later preferred to take ships across the Mediterranean), infantry—armed with bows, spears, swords, slings, and maces—walked. By far the most famous overwhelmingly walking adventure came at the beginning of the first wave, when Peter the Hermit led the undisciplined (and murderous) People's Crusade whose main body left Cologne on April 20, 1096, reached Niš in southern Serbia on July 3, Sofia on July 12, and Constantinople on August 1, covering a distance of 2,500 kilometers in 102 days. That meant averaging about 24 km/day—or, with just 6 hours of walking a day (slowed down more by the undisciplined progression of an unruly crowd than by the many women and children in it), the expected speed of 4 km/h.[46]

The speed of better organized military crusades was much faster, as mounted troops were transported across the Mediterranean by Venetian ships, but (as I will explain later in this chapter) the unpredictability of those sailings made it nearly impossible to set any firm dates for arrival and for the onset of military campaigns. The speed of long-distance trade on land was limited by the capabilities of draft and pack animals. Throughout the Old World, oxen were the dominant draft animals harnessed to wooden wagons and carts on flat terrain; asses and mules were the preferred pack animals in hilly and mountainous regions. The Theodosian Code spelled out the maximum loads allowed on Rome's paved *cursus publicus* roads, ranging from 66 kilograms for the two-wheeled *birota* drawn by mules to 492 kilograms for the heaviest oxen-drawn *angaria*, the latter proceeding no faster than 2.5 km/h—a speed like that of a sedan chair with two porters carrying a single person.[47] Pack animals could be a bit faster, with lightly laden mules averaging 3–4 km/h.

But the longer the distance, the less meaningful it is to speak of an average speed. The best example of this reality is the travel on the Old World's longest land travel route, from the coast of the Black Sea (Tana, near today's Rostov) to Ganzhou in China's Gansu province—one of the ancient links between imperial China and Europe. According to a Florentine merchant, Francesco Balducci Pegolotti (who himself did not make the journey), during the 1330s the distance of about 5,500 kilometers—via Sarai on the Volga, Utrār on the lower Syr Darya (in today's Kazakhstan), and Almaliq (in today's Xinjiang)—took 205 days using a combination of ox carts and pack asses.[48] That would be about 27 km/day and only about 3.3 km/h for an 8-hour travel day, but such a consistent speed was impossible because of interruptions due to rests forced by inclement weather, changing animals, reloading supplies, and waiting for a new caravan (often with armed protection) to be formed.

And such common delays were far surpassed by sudden flare-ups of local and regional conflicts (intra- and inter-tribal wars; invasions) that could close a route for weeks, months, or even years. On their first journey to the East, Niccolò and Maffeo Polo (father and uncle of Marco, whom they took along on their second Asian trip described by Marco's dictated book) had to spend three years in Bukhara before they could continue their journey and reach Kublai Khan's capital (Khanbaliq, near today's Beijing) in 1266.[49] Perhaps the best—and certainly the right-order-of-magnitude—conclusion is that a medieval journey along the northern route between the Black Sea and Xi'an (China's ancient capital), be it for trade or as an embassy from and to rulers and popes, took two years for a round trip—that is, some 12,000 kilometers covered at a speed that averaged only about 16 kilometers a day, well below the 4 km/h norm that could be expected during continuous, uneventful travel.

In some parts of the premodern world there were two faster alternatives for long-distance travel on land. The first was made possible by building and maintaining networks of paved roads that allowed carts or wagons pulled by horses to move preferred goods as fast as was practicable. The Roman Empire's extensive *cursus publicus*—whose first link, the Via Appia, connected Rome with Capua 199 kilometers to the southeast—was the best example of this option.[50] In his model of travel in the Roman Empire, Walter Scheidel (a historian at Stanford) assumes these daily averages: 12 kilometers for ox carts, 20 kilometers for heavily loaded mules, 30 kilometers walking, 36 kilometers for private vehicles making necessary rest stops, 50 kilometers for accelerated journeys by private vehicles, 56 kilometers on horseback, and 67 kilometers for fast carriages.[51] Using Scheidel's travel averages, an ox cart would have taken 16.6 days to go from Rome to Capua, a private vehicle 5.5 days, and a fast carriage just 3 days. The 6th-century Roman historian Procopius, without specifying the travel mode, said: "Now the Appian Way is in length a journey of five days for an unencumbered traveler; for it extends from Rome to Capua."[52]

Horses

In all premodern Old World societies, horseback riding to deliver messages and mail was, when properly organized, by far the fastest mode of urgent travel and messaging, made possible by the combination of equine speed and endurance—that is, the maximum distance they could cover under favorable conditions in a day. Horse gaits and speeds were already explained in the previous chapter, but here I must consider the endurance of animals and riders. An experienced rider could do 50–60 kilometers a day on a single, strong, well-fed horse, and much longer distances could be covered by relay rides by organized messenger services that had stables along the often-traveled routes where the riders mounted fresh horses at regular intervals. Four testimonies from three different eras (two from antiquity, one each from the Middle Ages and the 19th century) supply credible details of such organizations and their operating speeds.

Herodotus famously described the Persian royal service (*angarium*) established under Darius I (522–486 BCE) as having "so many horses and men set at intervals, each man and horse appointed for a day's journey. These neither snow nor rain nor heat nor darkness of night prevents from accomplishing each one the task proposed to him, with the very utmost speed."[53] An abbreviated version of this ancient appraisal eventually became an informal motto of the US Postal Service—and, as far as reliability and speed are concerned, many of the service's critics admire the original arrangement more than the modern version.

With 2,699 kilometers between Susa and Sardis accomplished in nine days, the "utmost speed" of the *angarium* implies an average of 12.5 km/h, a mean that could be reliably accomplished by the switching of horses every 18–25 kilometers and with the animals trotting through heat and night. Obviously, the prevailing mean speed did not maximize the performance, but it was deliberately chosen in order to increase the service's reliability by minimizing the ever-present risks of injury and exhaustion in the hot climate.[54]

The Roman rapid messenger service was a close variant of the older

Persian arrangements. *Mutationes*, stations for changing horses, were set about 15 kilometers apart and had some 20 horses at disposal, and several *mutationes* were interposed between *mansiones*—stations where messengers (and travelers) could rest, sleep, and eat. The maxima for continuous horse relays on well-paved Roman *cursus publicus* could be 250 kilometers a day, which means that a message from Augusta Treverorum, today's Trier on the Moselle in Germany, could reach Rome (a distance of about 1,400 kilometers) in less than eight days in summer, while a horse-drawn wagon transporting people and goods covered 30–40 kilometers a day.[55]

Yam, the Mongolian messenger service, was established by Genghis Khan's successor, his third (recognized) son Ögedei, who saw it as one of the crowning achievements of his rule. As *The Secret History of the Mongols* (written shortly after Genghis Khan's death) put it, because envoys used to be "dispatched to roam throughout the kingdom" without any fixed routes, "would it not be best to select post-staff and relay riders from every thousand in every direction, set up post-stations in every settlement and dispatch the envoys via the post-stations . . . The proposal on the matter of everyone setting up post-stations is the rightest of all of them."[56] Later, the system was expanded to serve the needs of the Yuan (Mongol) dynasty.

Not a few of Marco Polo's statements in his *Million*—the account of his long journeys through Asia, written down by a fellow prison inmate—are questionable, but his account of the Yuan *yam* has been confirmed by other sources. The network consisted of well-built and furnished waystations at intervals of 40–70 kilometers, with 200–400 horses (supplied by imperial command from the nearby areas) ready for relay runs. By the end of the 13th century, under Kublai Khan (1260–1294), the entire system comprised more than 300,000 horses and 10,000 waystations. Messengers (often tied to horses with head- and body-bands to prevent falls) rode as fast as the terrain and torch lights allowed, and 320 to 400 kilometers could be covered in a day. That implies an average speed of 13.3–16.7 km/h—a taxing but not impossible performance.

More than five centuries later, the most famous American rapid messenger service, the Pony Express that linked Missouri with

California before the completion of the first telegraphs and railways, could not do any better. The service was active for only 18 months, from April 1860 to October 1861, and it ran between St. Joseph in Missouri and Sacramento in California, with boats used for the last leg to San Francisco.[57] On April 3, 1860, the first rider left St. Joseph at 19:15; Salt Lake City was reached on April 9 at 18:45; Carson City, Nevada, on April 12 at 14:30; and the last of 75 messengers arrived in San Francisco around midnight on April 14. That would, for a journey of 3,146 kilometers, prorate to 13.1 km/h because the typical intended speed of 15–16 km/h could not be maintained through the night and in mountainous terrain.

The service opened with 119 stations, 500 horses and 80 riders, with horses changed every 16 kilometers and riders every 120–160 kilometers, and the record delivery was made in March 1861 when Abraham Lincoln's inaugural address was delivered from Nebraska to California in just 7 days and 17 hours (an average speed of 17 km/h) compared to normal trips of 10 days. The same journey would have taken at least 25 days by the stagecoach scheduled weekly between St. Louis and San Francisco, and 30 days for postal deliveries using steamers between New York and Panama and then Panama and San Francisco and mules across the isthmus. Setting up and operating the Pony Express was expensive, and transportation engineer Wayne Cottrell has shown that it could not compete with the telegraph even if it had lowered its prices by two-thirds.[58] Predictably, the great adventure shut down on October 26, 1861, just two days after the completion of the first telegraph link. But that link, as well as the later (completed in 1869) railroad and the first transcontinental road, followed the Pony Express course.

As for the fastest-ever gallop by a single rider, Buffalo Bill (born William F. Cody) claimed in his memoirs an extraordinary feat when riding the Pony Express as a teenager: using 21 fresh horses to complete 518 kilometers in 21 hours and 40 minutes, at an average speed of nearly 24 km/h.[59] "My boy, you're a brick, and no mistake. That was a good run you made when you rode your own and Miller's routes, and I'll see that you get extra pay for it," said his stage agent. That speed would be about 10 kilometers faster than the normal pace of the *angarium*, *yam*, or Pony Express—and all done without any mention of sleep! Too fast to be true, and there is no proof that this happened.

William Cody was born in February 1846, and hence just 14 years old when the Pony Express began its service. He worked for the owners of the short-lived company as a messenger but there is no record of him being one of the actual mail carriers. The confirmed Pony Express speed record belongs to a 20-year-old, Robert Haslam, who in May 1860, after completing his 75-mile run, covered another 115 miles because the rider who was to take over refused to go, being afraid (and rightly so) of Paiute attacks on the road. After a brief rest, Haslam returned to his home station, covering 380 miles in less than 40 hours, averaging more than 15 km/h.[60]

As far as galloping horses go, Cody's memoir is notable for containing two engravings that only confirm how ignorant people were about the way horses run, even by the 1870s. "Bob Scott's famous coach ride" on page 122 and a smaller picture of a single rider on page 92 show horses with both of their front and hind legs fully extended (forward and backward), almost on the level with the animals' bellies.

Cody's book came out in 1879, and neither its author nor its publishers were aware that by that time indisputable evidence showed that no horse could ever run like that. The actual motion of a galloping horse had been captured in detail for the first time by Eadweard Muybridge (1830–1904), an Englishman who became known in the United States thanks to his large, silver-print photographs of the Yosemite Valley that he first exhibited in 1868.

His most famous accomplishment came after Leland Stanford—one of America's leading industrialists, the president of the Central Pacific Railroad, governor of California, founder of the eponymous university in Palo Alto, and, finally, until his death a long-serving US senator and a horse breeder—hired him to clarify, once and for all, the sequence of the horse gaits. Did horses have one foot always on the ground when trotting, and all feet off the ground when galloping, and if the latter was true, was it as pictured in famous paintings including Théodore Géricault's 1821 *Derby at Epsom* (now in the Louvre) that has both the front and the hind legs fully extended (much as in Cody's book)?

Impossible and actual: gallops depicted by Théodore Géricault's *Le Derby de 1821 à Epsom* and in Muybridge's 1878 series *Sallie Gardner at a Gallop*.

The uncertainty was unequivocally resolved only in June 1878, when Muybridge's photographs—results of his perfected chrono-photography of animal locomotion capturing all four gaits of a horse at the Palo Alto track in California—ended the long-lasting dispute about the exact sequence of leg-ground contact in horse motion. After several years of experiments and improvements, Muybridge had set up a special track (with a white-sheet background for the best contrast) at Stanford's Palo Alto farm, lined up thread-triggered glass-plate cameras (with a shutter speed as brief as 1/1,000th of a second) along it, and reproduced the captured images as sequential black silhouettes, as well as on a rotating disc of a zoopraxiscope, a device of his invention simulating motion.

Successful runs with one of Stanford's horses, Sallie Gardner, clearly showed the animal having all four hooves off the ground during the gallop (the third image of the sequence)—but, quite contrary to Géricault's (and Cody's) rendition, only when its legs were tucked beneath its body (front legs bent backwards, hind legs bent forward), just before pushing off with the hind legs.[61] I find it remarkable, if not incomprehensible, that after a millennia-long history of riding and observing horses the resolution of such a basic fact regarding their fastest gait had to wait until photographic evidence became available during the late 1870s. Of course, no less remarkable is the fact that we got the first commercially successful modern bicycle (equal wheels, diamond frame, chain-sprocket drive) only in 1885, the same year Benz, Daimler, and Maybach patented their first cars (or more accurately, motorized carriages).

Only the use of steam power on rails ended the speed primacy of horses on land. The quest for speed on rails had been delayed, mainly due to concerns about the risks of high-pressure engines, into the late 1820s.[62] As I will show in the next chapter, the average speed of the very first commercial locomotives, introduced in 1829 and 1830, matched the speed of a typical horse gallop (25–30 km/h), and by 1840 no horse could run for hours at the speeds that were being sustained for long stretches by the steadily improving steam engine designs for locomotives. Similarly, only steam engine propulsion ended the primacy of the fastest sailing ships on water.

Sailing

As with all premodern developments, advances in sailing were gradual and uneven. The most ancient square sails, common in dynastic Egypt and in Greece, as well as in imperial Rome, were mounted perpendicularly to the ship's long axis and worked well only with winds astern. For such ships, "average speed" is not a meaningful term, and the speed of ancient voyages must be separated according to the winds that prevailed en route. After sifting through every notable reference made to the duration of ship journeys in classical writings, Lionel Casson of New York University put the typical speed range at 7.4–11.1 km/h (4–6 knots) with the wind, and 3.7–4.6 km/h (2–2.5 knots) against it, and Julian Whitewright's re-evaluation of ancient Mediterranean rig performance ended up with similar speeds of 4–5 knots in favorable and just 2 knots in unfavorable winds, both for the ancient square sails and the later lateen (triangular) sails.[63] Even the upper bound of the first (with the wind) mean was no better than the equivalent of a very slow horse trot; the second one was merely the speed of slow walking.

Going with and against the prevailing winds made an enormous difference: Roman ships could make it from Rome (Ostia) to Alexandria in 10–13 days; the return journey could last 33–53 days, three to four times as long. Nothing illustrates the challenge of sailing against the wind better than the remarkable lack of progress made (despite improved sail and ship designs) during the next 18 centuries. When Napoleon deserted his Egyptian army in August 1799, it took him 47 days to sail from Alexandria to Fréjus on the Côte d'Azur, a decidedly sluggish Roman-like progress! And far more dangerous was the trip taken from Alexandria seven years later by François-René de Chateaubriand, one of the era's most famous writers, on his return from visiting the Middle East.

He boarded his ship in Alexandria on November 23, 1806, and three days later they had hardly advanced as they remained within sight of Pompey's Pillar (a column erected in 300 CE on a rocky hilltop in the city); they sighted Malta on December 24, but then the west-northwest wind drove the ship south of Lampedusa and "for eighteen days we remained off the eastern coast of the kingdom of Tunisia, caught between life and death."[64] That is 49 days of not-quite making it from Alexandria to Tunis—at a time when even the westward crossing of the Atlantic by packet ships usually took no more than 40 days, and the eastward, wind-driven, journey could be accomplished in just three weeks.

With better sails, maximum ship speeds got marginally faster and, more importantly, the vessels could sail closer to the wind. Ancient square-sailed ships on the Mediterranean could sail only with wind 30 degrees off their course; medieval ships could proceed slowly with the wind at 90 degrees off their intended direction. The only option to overcome this limit was to take the best possible angle and proceed by slow (and laborious) zigzagging. Eventually, asymmetrical sails capable of swiveling around their masts made sailing closer to the wind possible. The most advanced sail ship designs of the late 18th and early 19th centuries—including triangular, lug, sprit, and gaff sails—could sail even when the wind was blowing 45 degrees off their intended course.[65] The 1492 transatlantic crossing took 37 sailing days, with hourly speeds ranging from zero to about 15 km/h, but implying a still very slow average speed of about 6.9 km/h.

Endeavour, the ship James Cook commanded on his first global circumnavigation between 1768 and 1771, was a fully rigged bark (with nearly 2,800 square meters of sails) but its top speed was just 8 knots (close to 15 km/h).[66]

But it would be pointless to calculate its average speed during any of these prolonged voyages: long sailings during the centuries of European expansion to Africa, Asia, North America, and Australia were, much like long trips on land, interrupted by too many mishaps and necessary rests (for taking on water and food, often for ship repairs) to calculate any meaningful average speeds.

Moreover, faster sailing depended not only on having better ship hulls, better sails, and better means of navigation (the marine chronometer), but also on the better health of crews.[67] Until the mid-18th century, long-distance voyages had very high mortalities, mainly due to scurvy. This constraint was solved by taking on board sufficient supplies of foods high in vitamin C (sauerkraut, lemons) and augmenting them by fresh vegetables and fruits at ports of call. But even the fastest ships able to navigate accurately and manned by healthy crews could have their progress halted or delayed by the temporary lack of winds. There is no remedy for a wind-propelled vessel that finds itself becalmed, for hours or days, and by far the highest chance of this happening is when crossing the equatorial doldrums.

Within this intertropical convergence zone, extending about five degrees north and south of the equator, air circulates upwards, resulting in the frequent absence of surface winds and thus extended periods of waiting before sail ships can resume their journey. There is no shortage of notable instances of prolonged becalming, including Ferdinand Magellan's expedition to circumnavigate the Earth: shortly after his ships left Sanlúcar de Barrameda on September 20, 1519, they were becalmed for 20 days off the African coast (today's Sierra Leone). But strong contrary winds could cause similar delays: even by 1800, English ships sometimes had to wait for many days before the right wind allowed them to enter Plymouth Sound, unable to sail against the prevailing westerlies and northerlies.[68]

Not only were ship journals full of repeated complaints about unbearably stifling calms, but the experience was vividly noted in Samuel Taylor Coleridge's great 1798 classic, *The Rime of the Ancient Mariner*:

> Day after day, day after day,
> We stuck, nor breath nor motion
> As idle as a painted ship
> Upon a painted ocean.[69]

The fastest sail ships of the mid-19th century could not avoid these spells of no motion. When Matthew Maury, an American ship captain and a pioneering climatologist, published his *Physical Geography of the Sea* in 1855, he noted how ships carrying emigrants from the UK to Australia were stopped by doldrums and "often baffled in it for two or three weeks; then the children and the passengers who are of delicate health suffer most. It is a frightful graveyard on the wayside to that golden land"—and provided at least a partial remedy when he published the first detailed compilation of wind directions and wind speeds gathered from a large collection of ship journals.[70]

Between 1750 and 1850, during the last 100 years of their dominance, sail ships did travel faster than before, but only in specific conditions. Daily observations of longitude and latitude in old ship journals make it possible to calculate daily "course made good," and large-scale analysis of such entries from British and Dutch ships shows that average sailing speeds rose by about a third, from 4.5 to 6 knots—but only in the moderate breezes (11–16 knots) prevailing during the North Atlantic summers.[71] Ships sailed faster in light breezes (4–6 knots) in 1830 than they did in moderate breezes in 1750, but in light breezes there were only minimal or no gains for both the Dutch frigates and the Royal Navy ships, and a slight decline for the Dutch East India Company ships.

The copper plating of ship hulls below the water line was the single most important contributor to this gain: between the 1770s and the 1820s, the copper sheathing of East India Company ships increased sailing speed by about 11 percent, making it possible to eliminate a stop at the Cape and cutting the total trip by 25–33 percent.[72] But frequent mishaps, necessary stopovers, and doldrums lowered average speeds to a crawl. Complete records of Dutch sailings to Dejima, the tiny island in the Nagasaki harbor to which Tokugawa Japan restricted foreign traders, showed that during the 18th century the trips from Amsterdam to Batavia (now Jakarta) averaged 245 days and that another month was needed to reach Japan, implying an average speed of less than 5 km/h![73] Add the attendant perils of poor nutrition and tropical diseases, and it is not surprising that about 15 percent of people who boarded Dutch ships during the 17th and 18th centuries died before reaching Batavia.

How much faster were the fastest 19th-century designs before they were made obsolete by steam engines? All record-breaking ships belonged to a remarkable class of sailing vessels: clippers.[74] These were merchant ships built mostly in American and British shipyards, with peak launching years between 1840 and 1860. They were specifically designed for speed, with narrow, low, pointed hulls, and carried limited high-value cargo and a small number of passengers, mostly between the US East Coast and California, from the US and UK to China, and from the UK to Australia. The fastest ones—three-masted fully rigged clippers—could hoist more than 30 sails whose total area would be on the order of 3,500 square meters.

Obviously, record speeds differed according to the distances under consideration. Donald McKay's *Flying Cloud* earned the record for the fastest sailing from New York to San Francisco around Cape Horn in 1854, completing the journey in 89 days and 5 hours, averaging 11.4 km/h (nearly 3.2 m/s) over the distance of more than 24,000 kilometers.[75] In 1851, soon after its launching, the ship sailed 1,311 kilometers in two consecutive days, averaging 27.3 km/h (7.6 m/s); and its one-day record was 687.9 kilometers, or nearly 8 m/s.

In 1854 *Champion of the Seas* (also McKay's design and build) logged an astonishing distance of 861 kilometers in 24 hours, averaging 35.9 km/h (nearly 10 m/s); while the fastest recorded short stretch belongs to McKay's *Sovereign of the Seas* at 41 km/h (11.4 m/s) in 1853.[76]

These designs for speed were so outstanding that *Flying Cloud*'s New York–San Francisco record run was bested only after 135 years of attempts by at least 250 vessels: in February 1989, *Thursday's Child*, a racing yacht crewed by three people, made it in 80 days and 20 hours.[77] But the comparison between a full-size commercial wooden ship that could carry more than 1,500 tons of cargo and was guided by traditional means, and a high-tech design of a comparatively small racing yacht made of advanced materials and guided by satellite navigation, is not even one between the proverbial apples and oranges; it is one of two (materially and logistically) different worlds! As a result, this conclusion is indisputable: mid-19th-century clippers were the fastest-ever sail ships—admirable designs that were as fast as a wooden wind-propelled vessel could go.

That those massive container vessels, some carrying more than 20,000 steel boxes filled with manufactures from East Asia to the Americas, Middle East, and Europe, can now travel faster than the maximum clipper speeds (typically at 24 knots, that is 44.4 km/h) is due to the power of modern internal combustion engines rather than to more ingenious vessel designs. Ships built by Donald McKay and his fellow clipper builders reached the pinnacle of sailing speeds, and when those giant container ships, powered by the world's largest diesel engines, slow down to "slow steaming" speeds (at 12–19 knots, an anachronic but common term in shipping) to save fuel on long intercontinental voyages, they go slower than the usual speeds of American clippers of the 1850s (14–17 knots)![78]

Collective speeds: the growth of populations, economies, and cities

Reconstructions of prehistoric population change are nothing but more-or-less arguable approximations. There is evidence indicating that the populations of the first-settled cultivators did not grow any faster than those of contemporaneous foragers, and that with annual rates on the order of a few hundredths of a percentage a year those populations stagnated for millennia.[79] But there is also evidence for relatively much faster population growth following the adoption of agriculture, even as the annual rates remained well below 0.1 percent—that is, a doubling speed greater than 1,000 years.[80] Many specifics throughout most of the historic era (the last five millennia) remain uncertain, but once we get information about the size of settlements and cultivated areas—and, later, actual numbers from (albeit partial, often counting only adult men for taxation or military needs) population censuses—we are on more reliable ground.

Absolute numbers for antique and medieval populations still have a rather wide range of uncertainty, but deducing the rates of population change carries lower errors. China's unique record of early censuses indicates an increase from about 60 million at the beginning of the common era to 100 million by the 11th century CE, implying an annual growth of less than 0.05 percent.[81] There are many estimates of global population totals going back to prehistory—and, as expected, their divergences decline with time.[82] We can be confident that global population growth did not rise above 0.05 percent a year during the first millennium of the common era (a doubling speed of at least 1,400 years), but by 1500 it had, most likely, doubled to 0.1 percent a year. Then it doubled again by 1600, declined during the 17th century, but recovered to about 0.2 percent during the next one. In 1800, by the end of the early modern era, the doubling speed of the global population was thus about 350 years, still too slow to result in substantial changes during a single (at that time much shorter!) lifetime.

During the latter half of the 18th century, a girl born in a small settlement of 100 people in Poland or Punjab would see—even if her village had remained unaffected by frequent military campaigns (during that time in Europe they involved all major powers, from Russia, Sweden, and Poland to Austria, Germany, and France) or had evaded yet another famine (India had three of them during the last two decades of the 18th century)—only an additional 10 villagers by the time of her death five decades later. That speed of population change, with two people added to a village of 100 peasants per decade, was close to a long-lasting no-change equilibrium, which resulted, in the absence of external events, from the combination of high birth rates and nearly equally high death rates.

Moreover, for many nations and regions, these centuries of stagnation were repeatedly interspersed with periods of major or even drastic population declines. Again, while absolute numbers remain uncertain and specifics arguable, there are no doubts about the sometimes astonishing rates of population loss. Many villages and small towns were destroyed and large shares of the populations of many cities killed (in sieges or after conquest) by advancing or retreating armies, be they Roman or Muslim, Crusaders in Anatolia or Syria, Mongol horsemen fanning out from the Pontic steppe during the 13th century, or massed movements of forces during Europe's religious and imperial wars of the 17th and 18th centuries.

Mass killings and the destruction of settlements, crops, animals, mills, forges, workshops, and material stores were reflected in severe

Jacques Callot's *Hanging*, from the 1633 cycle *Les Grandes Misères de la Guerre*.

and sometimes protracted economic declines that further reduced the already low rates of long-run economic development. As is the case with reconstructing population trajectories, quantifying long-term economic growth of the ancient, medieval, and early modern worlds is beset by numerous uncertainties, but historical economists have been no less willing than historical demographers to look for any shred of useful evidence (and, I should add, more likely to fill in the gaps with some truly heroic assumptions) in order to offer some quantitative conclusions.

Angus Maddison's *Contours of the World Economy, 1–2030 ad* has been the most quoted (but not always the most accurate) source of long-term economic growth rates.[83] According to Maddison, the world of the first millennium of the common era saw no discernible growth. The average annual growth rate of 0.01 percent implies the addition of a single unit to 10,000 units of measured output: at the end of another year would you notice an additional amphora or one more goat among 10,000 amphoras or goats? That growth also implies that the doubling of the economic product would take 7,000 years, starting at the beginning of the common era, a speed so slow as to be completely indiscernible by the most acute brain.

Predictably, that global mean hides substantial regional differences, and published estimates range from the high of 0.07 percent/year for Latin America (the rate obviously based on little but a concatenation of weakly based assumptions, as we have no written evidence from any part of that region for that period) to −0.03 percent for what are now 12 countries of Western and Central Europe (including the UK, France, Germany, and Italy but excluding Spain). That European rate would mean that the region's economic product was about 25 percent lower in the year 1000 than it was at the beginning of the common era (the time of the early Roman Empire), obviously a notable decline when seen in the millennial perspective but, once again, almost impossible to discern on an annual, or even decadal, basis by the people who lived during that time of very slow economic retreat.

Historians have also looked closer at the economic products of some major nations during different periods of the second millennium of the common era, and they have come up with more defensible conclusions based at least partially on a variety of relevant quantitative evidence. Not surprisingly, these reconstructed national accounts—reflecting the realities of economies dependent on low-productivity grain cropping, on human and animal muscles, and on slow speeds of work and travel—reveal either the complete lack of any long-term economic growth, or annual growth rates so low that they did not translate into any discernible year-to-year gains in wages, material well-being, or overall quality of life.[84] Spain did not show any economic growth between 1270 and 1850, and neither did central and northern Italy between 1300 and 1913, while the English economy grew by an average of 0.03 percent a year between 1270 and 1663.

Again, given short medieval and early modern lifespans, that was a speed of economic expansion that nobody could perceive on an annual basis, not even after a lifetime when (after 50 years) the gain would have amounted to just 1.5 percent of whatever was being counted (wages, articles of clothing, amount of meat consumed). A higher average rate of economic growth was calculated for imperial China: 0.1 percent a year between 1020 and 1850—but, yet again, this effect would have been barely noticeable in a lifetime (an exponential growth of just above 5 percent in 50 years), and even after eight centuries the gain would have amounted to a bit more than doubling the original value. In such stagnant economies there could be no concern about the speed of material and income gains: in good times there were simply no noticeable improvements; after natural disasters or violent conflicts came long periods of hardship (economic and population retreat) and the eventual slow restoration to previous levels.

Additional evidence of long-term stagnation, or of growth so slow that it made notable differences only after centuries and could not be discerned on a year-to-year or decade-to-decade basis, comes with crop yields and cathedrals. Cultivation of the same staple grain, legume, oil, and tuber crops by using the same draft animal harnessed to only mildly improved wooden plows, seeding done simply by broadcasting, and harvesting by sickles or scythes resulted in the centuries-long stagnation of yields. English wheat yield in 1600 differed little from harvests in 1200—or from American and French yields in 1820![85]

Many large construction projects depended on building techniques that had not changed for centuries. Ox-drawn wagons delivered large stones to building sites; wooden cranes (powered by men walking inside large wheels or by harnessed animals) were used to lift stone and bricks to high vaults; big pieces of timber were cut and assembled into massive scaffoldings needed to erect large cupolas; and bricks or stones were laid down by masons with hods and trowels. This was inherently slow construction, often further delayed by frequent financing problems and interruptions caused by wars. As a result, many projects took centuries to complete, with the great European cathedrals providing many examples: York Minster, 252 years; the Duomo di Milano, 579 years; St. Vitus in Prague, 585 years; Cologne Cathedral, 632 years.[86]

Not just large buildings, but large cities were very slow to emerge. A city has always been a product of population and economic growth, and given how mutually reinforcing those two variables have been, we should see the term "product" not only as a descriptor of an outcome of their interaction but also in its algebraic meaning—as their multiplier. This simple characterization explains why cities in general, and large cities in particular, housed only very small shares of a society's population during the premodern eras: no or minimal long-run population growth, the necessity to have most people engaged in low-yield food production powered by animate energies, and nonexistent or negligible rates of economic expansion could not result in anything but societies where more than 90 percent of people lived in villages, and where only a few very large cities emerged as centers of far-flung empires.

Reconstructing the long-term global speed of urbanization is thus a trivial exercise for about 95 percent of recorded history—merely plotting a flat line that does not rise above 5 percent of the world population count until the 18th century, and that is still only at about 7 percent by 1800.[87] Those rare large cities that were the first to pass the million mark had starring roles in history: imperial Rome of the 2nd century of the common era, the Tang dynasty's Xi'an, and the Abbasid caliphate's Baghdad of the 9th century. But subsequently, even in Europe and China, the speed of urbanization slowed down, with a revival coming only after 1600.

By the end of the early modern era, there were regions where various gradual improvements resulted in notable gains that set such societies above the prevailing means. England and Holland in the late 18th century were the two foremost European examples where economic growth—based on higher crop yields resulting from more intensive farming, on the mass deployment of wind power, on incipient industrialization, and on extensive foreign trade—became a readily discernible presence changing the two societies at a faster pace than anywhere else.[88] But by the early 19th century, when Austria's Metternich and France's Talleyrand agreed on new continental arrangements in Vienna in 1815, most (and by that I mean more than 95 percent) of the world still resembled more the early 17th century or even the early 16th century than it did the early 20th century. However, the time of great acceleration was about to begin.

But before turning to that new age of speed, I would be remiss if I did not mention one notable gain that took place during this era of stagnation and very slow growth: the much-increased speed and even more increased power of destruction. For millennia, most combat was hand-to-hand, with all kinds of edged or blunt weapons, and arrows (from bows or crossbows) were the only way to project destructive power across moderate (40–200-meter) distances. The kinetic energy of these thrusts (as fast as 10 m/s) or arrows (flying speeds up to 40 m/s) could inflict mortal wounds to unprotected bodies, but in absolute terms it was not high: 15–20 joules for a flying arrow; 200 joules for an expertly wielded *katana* (Japanese sword) cut.[89] Roman *ballista* could throw a 25-kilogram stone at the speed of some 50 m/s, giving the projectile kinetic energy of some 30,000 joules.[90]

In comparison, a typical 38-gram leaden bullet from a musket used during the Thirty Years' War (1618–1648) had a speed of 500 m/s and hence a kinetic energy of about 4,700 joules, while an iron ball from an 18th-century cannon had a kinetic energy of 300,000 joules.[91] This means that, by the early modern era, the kinetic energy of personal weapons had increased two orders of magnitude and that of heavy projectiles by an order of magnitude—enormous gains compared to the stagnation or very slow growth of the dominant prime movers (human and animal muscles) and the technical capabilities and speeds deployed in production, construction, and transportation. As a result, the instantaneous speed of killing increased from a single man succumbing to a mortal sword, axe, or arrow wound to dozens or even scores of people in a group hit by an iron cannonball. But this needs to be put into a wider perspective.

Gains in destructive capabilities during the early modern era were an exceptional component of military developments, as the progress of large invading armies remained fundamentally dependent on large ox trains. Smaller units could live off the land (by requisitioning food and by pillaging), but an army of tens or hundreds of thousands of men moving through thinly inhabited, wooded, and boggy regions had to carry its supplies. In his memoirs of the disastrous Russian campaign, Philippe-Paul de Ségur, one of Napoleon's top generals, recalled how Prussia alone had been forced to contribute 44,000 oxen and 3,600 wagons with harnesses and drivers to the invasion. Consequently, it is not surprising that he enumerated the initial force at 490,000 men, 1,372 pieces of cannon and "innumerable herds of oxen"—but even so the Grande Armée was short of draft animals from the very outset.[92] Hence the outcome: in an encounter of unprecedented intensity, during the Battle of Borodino (September 7, 1812) the French and Russian artilleries fired 110,000 rounds in total—or three cannon shots a second.

But after crossing the Neman River it took the invaders 76 days to reach the battlefield, and that was advancing 11.6 km/day or (with daily 8-hour marches) merely 1.5 km/h, the speed of heavily laden oxen-drawn wagons on miserable roads.[93] Oxen imposed a fundamental speed limit, and a painfully slow advance was ended by rapid slaughter.

As for those relatively large gains of destructive speed and energy that were attained during the last two centuries of the early modern era, they were soon reduced to minor achievements on the road to modernity. During the 19th century, technical advances and social transformations brought unprecedented speed gains across the entire economic spectrum, from agriculture to warfare, and the 20th century built on that admirable foundation to normalize even greater speeds. And, for the first time in history, on a truly global scale. These universal accelerations will be the focus of this book's penultimate chapter, as we move from the world of metabolism-restricted and muscle-limited animate speeds (marginally aided by flowing water and wind breezes) to a world whose traveling speeds are dominated by the rotations of fossil-fueled machines and electric motors and whose communications now proceed largely at the ultimate speed— that of light. Get ready for some dizzying speeds!

4. The Fast Lane

How modern societies accelerate

The widely accepted periodization of history labels the three centuries preceding 1800 as the world's early modern era, but as I explained in the previous chapter, during that period practical steps toward modernity remained limited due to the continued dominance of animal and human muscles, in some regions supplemented by the contributions of windmills and water mills. Moreover, even in the most advanced countries, pre-1850 gains remained modest in comparison to the unprecedented acceleration of just about everything that began to unfold during the latter half of the 19th century, and continued to speed up during most of the 20th century.

Removing the restrictions that were imposed by animate prime movers on the accomplishments of agricultural and industrial production, as well as on the speed of transportation, led, commonly, to order-of-magnitude gains in the speed of agricultural and industrial tasks, eventually to even larger gains in the speed of intercontinental travel and mass-scale industrial production, and to even more stunning gains in the speed of communication and the mass-scale diffusion of information. I will trace these developments by looking at the history of faster rotations and their multifaceted impacts on the modern world, and, finally, by taking a closer look at the consequences of previously unthinkable (essentially instant) communication and information flows.

External (steam) and internal combustion engines (fueled by liquids or gases) enabled conversions of the chemical energy in fossil (and synthetic) fuels—as well as in ammonia and hydrogen—to thermal and kinetic energy, and electric motors have made it possible to turn electricity into kinetic energy. As different as these machines are in terms of function and conversion efficiency, they share a key attribute: their operations end up producing rotational kinetic energy.

Those rotations are then deployed for powering land-, water-, and airborne transportation as well as a still-growing array of industrial and household tools and machines, and (a curiously underappreciated contribution) electric motors are used to generate the high speeds of the metal machining required to turn out an enormous array of industrial and consumer products.

From planks of wood to jetliners

Most of the work done in traditional societies involved linear motions—that is, changes of the position of an object in a straight (rectilinear) or a curving (curvilinear) direction—made possible by the exertions of human and animal muscles. The premodern world was suffused with such actions: men plowing fields with moldboard plows pulled by harnessed animals, or scything ripe crops; loads carried on the backs of porters or pack animals; women walking while balancing bundles of fuel wood or clay jars on their heads; river or canal barges floating downstream or pulled by oxen; large stone columns or blocks dragged by teams of workers on building sites of temples and cathedrals; bundles of bricks or stones pulled up on construction sites by ropes slung over pulleys. Ships sailing to nearby ports or to distant continents were the only category of common linear (or more often curvilinear or zigzagging) motion not powered by animate energies.

Reciprocating motion is a version of linear work: it involves repetitive, often tiresome, back-and-forth or up-and-down movements of an object. Manual grinding of grain on a flat or a curved quernstone, hand-sawing wood, polishing glass and metal (including swords) on simple benches, lifting water using a piston hand pump, and ringing heavy church bells were common examples of such endeavors—as was the rowing of small vessels, or very large galleys propelled by up to 180 seated men.[1]

One of the most demanding reciprocating exertions in many Old World societies was making planks from large tree trunks by using sawpits (or elevated arrangements), with one man standing above the

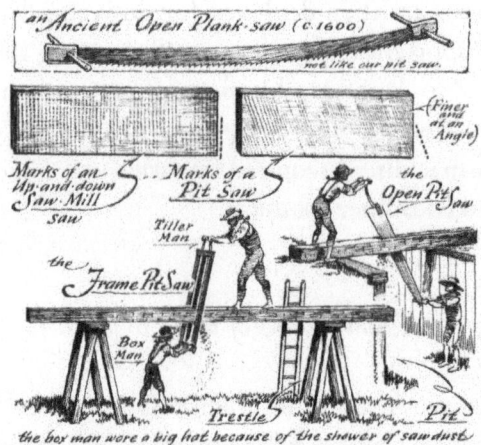

Before waterwheels and windmills took over, reciprocating exertions were the only way to cut—slowly and exhaustingly—large pieces of lumber.

log and the other one in the pit below it, as they alternated in pulling a long two-handled saw.

Another category of reciprocating motion, percussion (cable) drilling, was practiced in China since the Han dynasty (ancient Rome's contemporary), and two millennia later it launched the modern oil and gas industry.[2] A heavy metal hammering bit is attached to a long cable (rope, wire) supported on a tripod. Lifting and dropping the bit breaks down the soil or rock in the borehole, and the debris is extracted with a bailer.

The Han Chinese employed this technique first in Sichuan to drill for natural gas (burned to evaporate brine to produce salt in the land-locked province), using bamboo cable and derricks and heavy iron bits, with the reciprocating motion actuated by 2–6 men jumping on a lever; alternatively, a group of men could power the process by pulling the cable over a pulley and then letting the bit fall.

This simple process eventually produced some deep boreholes in China, more than 100 meters by the 10th century and a record depth of 1 kilometer in 1835. The discovery of oil in Pennsylvania in 1859 relied on the mechanized version of the same technique, with small steam engines, rather than human exertion, powering the reciprocating motion. Although a superior way of drilling deep wells—using rotary drill bits and also rotary tables—was introduced after 1900, mechanized percussion drilling dominated all well drilling until the mid-20th century.

Tasks accomplished in traditional societies by deploying muscle power in rotary motion remained relatively limited. Spinning wheels were unknown during antiquity; they made their way from India (where they appeared around 500 CE) to Europe only during the High Middle Ages (1000–1300). Similarly, a crank—an arm attached at a right angle to a rotating shaft, hence transforming its rotation to reciprocating movement—was first used in Han dynasty China but it appeared in Europe only a millennium later, by the 9th century. Horizontal windlasses (winches), vertical capstans, and wooden treadmills were used in mining, construction, and manufacturing, but their usefulness was limited by the manpower (or harnessed animals) used to rotate them.[3]

Turning the wheel

Linear (or curvilinear) animate motion was used to pull wheeled vehicles (wagons, carts, carriages) or to push wheelbarrows (with a centrally mounted wheel used in China since the Han dynasty, and a front wheel in Europe since the 12th century CE), but their premodern use was limited not only by animal and human metabolism but by the poor state of roads and poor wheel design (usually with unlubricated wooden fixed or rotating axles).[4] A potter's wheel (likely a Sumerian invention, eventually used throughout the Old World but never reaching the pre-Columbian Americas) has been in use for more than 5,000 years, but hand- or foot-power limits it to slow rotations.[5]

Consequently, waterwheels and windmills were the most impressive (and the most powerful) examples of machines converting natural energy flows into rotary motion. But, again, for many centuries the heavy designs of these machines limited their power and made for slow operation. The rotation speed of premodern wooden waterwheels was as slow as 1.5–2 revolutions per minute, which means that the outer rim of a large (about 5-meter-radius) undershot wheel rotated at a speed no higher than 1 m/s; overshot wheels moved faster—one with a 3-meter radius would run at 10 rpm, or about 3 m/s. The typical speed of large (10-meter-radius sails) Dutch windmills used to grind grain was 50 rpm, which means that the sail tips had a speed of 50 m/s, making these machines the most powerful energy converters of the preindustrial era.[6] But, as I will soon note, that speed (about five times faster than the record 100-meter run) appears minuscule when compared to the speed of blades turning large modern wind turbines.

Steam engines—converting the chemical energy of a fossil fuel to the kinetic energy of steam that moves pistons—were the first machines that opened up the possibility of unprecedented increases in useful power. Thomas Newcomen's first heavy, non-condensing (and hence very inefficient) engines delivered only reciprocating motion. They were in use from 1712 and worked with 12–14 strokes per minute, while James Watt's improved designs (with a separate condenser and other design improvements) averaged 20 strokes per minute.[7] These new powerful engines could be used directly for water pumping from mines, but their widespread adoption depended on the use of Murdoch's and Watt's famous sun and planet gear (a pair of linked cog wheels, one running around the other) that transformed Watt's beam engine's 20 strokes per minute to around 40 rotations per minute.[8]

But a simple crank—an ancient Chinese invention, initially for winding silk—and arrangements of components ensuring constant speed rotation (starting with a flywheel) were all that was needed to enable this transformation that allowed steam engines to power countless industrial tasks and enabled the first mode of rapid, reliable, and inexpensive land transportation. The most common place to see large cranks (activated by horizontally placed pistons) during the age of steam was by looking at the large driving wheels of any steam locomotive.[9]

One of the first large-scale commercial applications of this new fast rotary power left us with a memorable phrase. In 1786, Albion Mills in London's Southwark became the first commercial flour mill in the world operated by steam power.[10] The mill was powered by Watt's (for that time) large 50-horsepower engine that was able to rotate 20 pairs of millstones, each producing nine bushels (that is, about 170 kilograms) of flour per hour, with enough power to spare for mechanical fanning, sifting, packing, and loading into barges. This made all existing mills uncompetitive, and on March 2, 1791, the mill was destroyed by fire.

A contemporary engraving depicted three independent millers (one lifting a toy windmill) as "merry mealmongers" dancing against the background of the burning mill, wishing "Success to the mills of Albion but no Albion Mills."[11] The mill's opponents had referred to the Albion Mills as a satanic invention, and it is quite likely that this impressed William Blake (he lived close by) and led him to write a poem whose often-quoted line asks: "And was Jerusalem builded here / Among these dark Satanic Mills?"[12] The reference has been overused when describing the miseries of England's industrialization, but its origin does not appear to be in iron works—a more obvious site for heat, flames, and grime—but in the milling of wheat!

How quickly have we progressed?

The shift from muscles, water, and wind took longer than schoolbook histories of the 19th century would lead you to believe, and modern historical research offers rich correctives to the simple, rapid, and widespread "Industrial Revolution" narrative. This simplification of a complex and extended process can be illustrated on many levels. Steam engines were rapidly adopted only by some industries (textile mills, sawmills, ironmaking) and the British economy remained largely traditional until 1850; and hence, as historian Albert Musson put it, "the typical British worker in the mid-nineteenth century was not a machine-operator in a factory but still a traditional craftsman or labourer or domestic servant."[13]

Similarly, US data show that in 1850 Massachusetts had 1,268 enterprises powered by water and just 73 powered by steam (a mere 6 percent of the total), and according to labor economist Carroll

Daugherty's detailed reconstruction of all installed US power, in 1869 steam engines accounted for about 33 percent of the total, water- and windmills for about 8 percent, and draft animals for 59 percent![14] During the first half of the 19th century, steam power's most revolutionary application was in land and water transportation, and both of these required the development of high-pressure steam engines whose reciprocating power could be converted (with minimal losses) to appropriately rapid rotations.

Land applications came first, and in 1829 Stephenson's famous *Rocket* emerged victorious in the first trial of locomotives to decide which machine to choose for the world's first commercial railway.[15]

Bigger and more powerful locomotives followed, and new designs were further advanced by the expansion of railways in the United States, where intercity distances—measured in hundreds of kilometers—called for more powerful and more efficient machines. Limits to locomotive speed were set primarily by the frequency of strokes. At 150 km/h, a locomotive's driving wheels make six revolutions per second (360 rpm). Obviously, this performance has its limits. During the 1920s, the Northern Pacific Railway's powerful 4-8-4 locomotives (with four leading and four trailing wheels) had eight 70-inch diameter driving wheels turning at 533 rpm or 8.9 revolutions per second, enough to achieve nearly 179 km/h—perhaps the highest speed documented in regular passenger service.[16]

Much lower speed limits apply to steam engine–powered marine propulsion. The first steamships of the 1830s and 1840s had paddlewheels, emplaced aft or mid-hull on both sides and rotating at less than 50 rpm.[17] The first two propeller patents were filed independently in 1836 by Francis Pettit Smith ("for an improved propeller for steam and other vessels") and John Ericsson ("for an improved propeller applicable to steam navigation").[18] In 1839, *Archimedes* became the first steamship (still with sailing masts) to be powered by a screw propeller, and the new mode of propulsion quickly conquered the market.[19] Many improvements were made (to prevent the outward flow of water, and to reduce vibration), and for the next 70 years the propellers kept getting larger—because the larger the propeller diameter, the higher the propeller efficiency and the lower the

Rotary motion in ship propulsion: *Olympic*'s three propellers.

optimum propeller speed. *Olympic*, sister ship of the more famous *Titanic*, had two three-bladed manganese bronze wing propellers with a diameter of 716 centimeters (23.5 feet) and a four-blade central propeller (driven by a small low-pressure steam turbine): when they turned at 75 rpm, the ship's speed was 22 knots (40.7 km/h).[20]

The speed limits of propeller propulsion are dictated by the increasing perils of cavitation.[21] The faster a propeller rotates, the greater the difference between the pressures created on the attack and suction side of its blades. As increased speed leads to lower pressure on the suction side, water can start evaporating at ambient temperature, and bubbles form, grow, and eventually implode—creating high-pressure microjets whose impact erodes the propeller blades. For speeds of less than 35 knots (about 65 km/h) the problem can be minimized by using multiblade screws with a large diameter and wider blades, but in faster ships cavitation is almost inevitable.

As trains and new steamships were making the news by breaking speed records during the second half of the 19th century, the conquest of industrial production by steam engines also proceeded rapidly, and by 1880 steam-powered rotary motion had mechanized and sped up the output of every industrial sector. By 1900, many large stationary engines worked with speeds ranging from 120 rpm for large (1,000-horsepower) machines used in new electricity-generating plants, to

400 rpm for small (30-horsepower) industrial models.[22] There was a concurrent rise in efficiency from less than 1 percent for Newcomen's machines to more than 4 percent for Watt's later designs, to about 10 percent for the best commercial models of the mid-19th century, and, finally, to a record performance of about 15 percent by pre–First World War machines.[23]

But before steam engines could become universally dominant, three innovations of the 1880s—steam turbines invented by Charles Parsons, lightweight external combustion engines powered by liquid fuels, and electric motors—brought about their eventual demise. The steam turbine is a highly efficient rotary machine, and Parsons's first prototype built in 1884 had a power of just 7.5 kilowatts (equivalent to about 10 horses) but it ran at a dizzying speed of 18,000 rpm, with blade top velocities reaching 76 m/s— too fast for any useful contemporary applications.[24] But the new technique matured rapidly: by 1891, Parsons's first 100-kilowatt turbine, installed in Cambridge, produced electricity by rotating the generator at 4,800 rpm.[25] During the 20th century, the highest capacities of steam turbines rose by four orders of magnitude (to more than 1 gigawatt), and the largest machines (directly coupled with generators) now operate at 3,000 rpm.[26]

In contrast, two- or four-stroke internal combustion engines burning liquid refined fuels and now dominating road (and off-road) transportation and shipping are, much like steam engines, reciprocating machines—and crankshafts are needed to convert the up-and-down motion into the rotation needed for the transmission, driveshaft, and wheels or propellers.[27] This comes, inevitably, with efficiency loss but it provides easily controllable power for both stationary and mobile applications, and desirable acceleration for vehicles and ships. The first vehicles powered by internal combustion engines appeared during the 1880s, their use accelerated after 1900 (Ford's Model T was released in 1908), and by 1919 their accumulated capacity dominated America's installed power: they accounted for 62 percent of the total, compared to 29 percent for steam turbines and engines, 6 percent for draft animals, and just 2 percent for water-wheels, water turbines, and windmills.[28]

New rotations

More than a decade after Carl Benz, Wilhelm Maybach, and Gottlieb Daimler built the world's first gasoline-powered internal combustion engines for their carriage-like road vehicles, Rudolf Diesel added another internal combustion option: an engine fueled by heavier liquid fuel, operating with a much higher compression ratio and hence without sparking.[29] The engine saw its first commercial pre–First World War applications in shipping; between the wars it began to gain market shares in trucking and for freight locomotives; and since 1950 it has dominated all forms of heavy shipping (from oil tankers to container vessels), heavy road and off-road machines, and in some countries it also became a favorite for cars.[30]

In total, the world now has about 1.5 billion road vehicles powered by gasoline and diesel engines whose reciprocating motion is smoothly transformed into wheel turns; and the world's largest diesel engines power high-capacity vessels ranging from oil tankers and bulk ore or grain carriers to container ships.[31]

Electric motors started as direct current (DC, battery-powered) machines even before the first commercial generation of electricity in 1882, but the way to their mass-scale adoption was opened only after Nikola Tesla's patent for a two-phase alternating current motor was bought in 1888 by George Westinghouse.[32] His first mass-produced (125-watt) AC motor powered a small fan, and the first three-phase AC motor (each phase offset by 120 degrees) was built by a Russian engineer, Mikhail Dolivo-Dobrovolsky, working in Germany.[33] Thanks to their versatility and high efficiency (at least 65 percent, and up to 95 percent) of converting electricity into rotary motion, electric motors are now among the world's most ubiquitous artifacts, running everything from high-speed trains to metal-cutting machines and from elevators in the tallest skyscrapers and cranes lifting steel containers to kitchen mixers and coffee grinders. The rise of modern electronics and battery-powered tools has also multiplied the uses of small DC motors. Obviously, given the enormous variety of these machines, their rotations range across several orders of magnitude.

In North America, with electricity supplied at a frequency of 60 hertz, small motors commonly used in home and office appliances and by many types of services, ranging from drills to air condition-ers to garage openers, have speeds of 900–3,600 rpm under no load (spinning freely) and 850–3,450 rpm under rated loads (that is, when working); in the EU (50 hertz), the latter range would be 700–2,850 rpm.[34] Motors in electric cars turn at about 15,000 rpm (compared to about 6,000 rpm for gasoline vehicles); Dyson's digital DC motor for small appliances operates at speeds up to 104,000 rpm; in 2008, ETH Zurich tested a miniature motor that exceeded 1,000,000 rpm; and Celeroton, a company that grew out of that research, sells a tiny (36-gram) DC motor that runs at 500,000 rpm.[35]

And the final (20th-century) addition to the new world of rotating prime movers was the gas turbine, with the first practical designs of the late 1930s used for experimental jet aircraft and for the generation of electricity.[36] By the late 1960s, jet engines (gas turbines in flight) dominated long-distance aviation, and more recently large stationary gas turbines have become the most efficient generators of electricity. The largest (twin-aisle) modern jetliners are powered by turbofan engines whose fan blades (visible inside their round cowlings) rotate at about 3,000 rpm, while the low-pressure shaft of the turbine (com-pressing the air that bypasses the engine) spins at 12,000 rpm and the high-pressure shaft (compressing the air fed into the combustor) rotates at about 20,000 rpm.[37]

Even if you do not pay attention, you encounter many of these rotary machines every day, and you can hear some of them making plenty of noise as they drill into the ground, grind coffee, or lift departing airplanes. And people in the downtowns of large cities can still often hear the rhythmic noise of linearly moving machines: diesel or hydraulic pile drivers installing the steel or concrete foundations of tall buildings.[38] But there is a large cat-egory of extraordinarily important rotary machines that has been impressively upgraded during the past 150 years but that—unless you are among their designers, builders, or operators—remains almost completely out of sight: the lathe has been used since antiquity to produce objects ranging from parts of furniture (chair legs) to spinning wheels, and from cups and bowls to sieves and pulleys.[39]

In the late 17th century, Joseph Moxon, a member of the Royal Society, described the lathe's working succinctly:

> any Substance, be it *Wood, Ivory, Brass,* &c. pitcht steddy upon two points (as on an *Axis*) and moved about on that *Axis*, also describes a Circle Concentrick to the *Axis*: And an Edge-Tool set steddy to that part of the outside of the aforesaid Substance that is nearest the *Axis*, will in a Circumvolution of that Substance, cut off all the parts of Substance that lies farther off the *Axis*, and make the outside of that Substance also Concentrick to the *Axis*. This is a brief Collection, and indeed the whole Sum of *Turning*.[40]

The simplest ancient designs relied on reciprocal linear motion: either a turner cutting an object and a helper pulling back and forth on a cord, or—with foot-treadle-operated medieval spring pole lathes—the turner cutting on the downstroke and a cord connecting the treadle with a spring pole providing the recoil motion. Continuous wheel-operated rotary lathes made it possible to work with heavier objects at faster and less variable speeds and with higher efficiency: Diderot's encyclopedia illustrates the state-of-the-art design of the mid-18th century.[41]

These rotary lathes could work comparatively fast with wood but only slowly with small pieces of the softest alloys such as brass or pewter. The first metal version to cut iron was made only in 1800, when Henry Maudslay designed a screw-cutting lathe with a clamped (rather than handheld) cutting tool able to produce highly accurate standardized machine parts.[42] Its many subsequent improvements and modifications (including steam-operated machines) allowed for mass-scale industrial production.

Not surprisingly, the lathe has earned the sobriquet of "the mother of all machines," as today's computer-controlled lathes produce, with astonishing precision, countless parts for other lathes and metal-working, electrical, combustion, and industrial machines from aircraft to tankers: screws, nuts, bolts, cylinders, axles, pistons, crankshafts, barrels, rods, sleeves, baffles, connectors. Metal machines making more parts for more metal machines is the quintessential way of the modern world, and yet another grand confirmation of the general conclusion that "the technical advance that characterizes specifically the modern age is that from reciprocating motions to rotary motions."[43]

Round and round: when rotating motions took over everything

Undoubtedly the most obvious category of speed (*sensu stricto*) gains has been the progress in transportation. Most of today's 8 billion people have experienced high speeds on land and water and in the air—both directly, as operators or passengers, and indirectly, from watching car races and rapid trains to noting jetliner contrails against the blue sky. These speeds must be measured against the premodern norms: as we have seen, they ranged from 4–8 km/h (1.1–2.2 m/s) for walking to about 15 km/h (about 4 m/s) for the fastest speeds on land (mounted messengers) and to similar typical speeds for sail ships (with the 19th-century clipper maxima roughly twice as fast). I will progress chronologically, starting with steam engines followed by machines powered by internal combustion engines (cars, ships, airplanes), and, finally, electric motors (installed in an enormous variety of industrial and consumer products, and powering, most notably, trains and cars).

In all cases there are several speeds to consider. At the top is the highest designed speed of a particular machine that could be achieved under optimum conditions: by steam locomotives on long, straight, and slightly downhill stretches; by gas turbine–powered aircraft at cruising altitudes with no headwinds. Next come typical average operating speeds. In industries (be they textile or grain mills, car assembly lines, or steel scrap-processing facilities), the speeds of engines and motors must ensure desired productivities and allow for unforeseen downtime and scheduled maintenance. Similar downtime and maintenance considerations apply in transportation.[44]

In addition, for airplanes the averages are lowered due to the slower speeds required to reach and descend from cruising altitude, and similar speed reductions apply to arriving trains and docking ships. In addition, gate-to-gate speeds in flying are often reduced due to congestion during departure or arrival (and in winter, also due to the need for de-icing).[45] When applied to modern flying, these speeds are: 900 km/h for designed maxima; 840 km/h for the most economical operation; about 750 km/h gate-to-gate with tailwinds; about 600 km/h gate-to-gate with headwinds; and often less than 500 km/h with frequent airport delays (and in the latter half of 2022, down to zero as large shares of North American flights were cancelled, often for several days).[46]

The speeds of railway steam engines progressed rapidly from a top speed of 48 km/h for Stephenson's *Rocket* in 1830 to a maximum of 100 km/h achievable briefly on short straight stretches with light loads by 1850. But typical average speeds on American railways (including stops) were less than 40 km/h (and only 20 km/h in mountainous terrain in the west) during the 1870s, around 60 km/h by 1900, and by the late 1920s the fastest intercity trains averaged up to about 100 km/h.[47] Then the new, powerful, and streamlined designs raised the records to new levels: in 1934, the LNER Class A3 locomotive *Flying Scotsman* broke 100 mph (160 km/h); in May 1936, Germany's highly streamlined Borsig locomotive reached 200.4 km/h during the Berlin–Hamburg run with a 197-ton train; and in July 1938 the LNER Class A4 *Mallard*, on a downhill run but with a heavier (240-ton) train, set the fastest steam locomotive record of 203 km/h (about 56 m/s).[48] Even during the 1950s, before they were replaced by diesels, the actual speeds of steam-powered trains on local runs (including frequent stops and curving tracks) remained on the order of 50–60 km/h.

Diesel-powered locomotives were used first for freight trains, and since the 1930s in passenger service. In 1934, shortly after its introduction, the streamlined silver *Zephyr* set a new long-distance speed record of 124 km/h (and short-term speed record of 181 km/h) as it traveled non-stop 1,633 kilometers between Denver and Chicago.[49] But the most important step in accelerating railroad transport was the switch from internal combustion engines to electric motors—the most efficient rotary convertors—which made modern rapid trains possible. The first electric trains date back to the 1890s, but steam and diesel dominated both long-distance passenger and freight traffic until the 1950s, and it was not Europe, with its tradition of electric trains, but post–Second World War Japan that took the first practical steps toward rapid rail.

This did not require any fundamental new inventions, just the meticulous application, improvement, and reliable integration of well-known techniques—including AC supply from overhead wires, aerodynamically styled light aluminum car bodies, regenerative brakes, and a ballasted track with concrete sleepers. By October 1, 1964, the Tōkaidō Shinkansen between Tokyo and Osaka—with speeds of 210 km/h for the faster *Hikari* trains—was in operation, and it was followed by extending the links to both the southernmost island and to the northernmost part of Honshu, as well as to Niigata on the Sea of Japan.[50] Tōkaidō speeds were increased to 220 km/h in 1986; in March 1992, the new 300-series *Nozomi* trains reached speeds of 270 km/h; and since 1997 Sanyō line trains (with their 15-meter-long aerodynamic nose and wing-shaped pantograph) have speeds of up to 300 km/h.[51]

The frequency of trains on the Tōkaidō line increased from 60 trains in 1964 to 285 by the year 2000 and to 365 by 2023, with 432 trains during the August Obon holidays—yet the average annual delay is a mere 24 seconds![52] During the first 50 years of its operation, the Shinkansen moved nearly 6 billion people without a single collision, without a single fatality, and with only two derailments (without injuries) caused, respectively, by an earthquake in 2004 and a blizzard in 2012. Japanese engineers also pioneered all the conditions needed for rapid rail: a dedicated track with no level crossings, large-diameter (2.5-kilometer) curves, a steepest gradient of 2 percent, prefabricated high-performance concrete sleepers, seamless rails, and automatic train controls.

France pioneered European rapid trains with the Paris–Lyon route in September 1981, and the 1990 line to Bordeaux was the world's first scheduled connection operating at up to 300 km/h (now common also in Spain and Italy).[53] Germany's ICE and Spain's AVE were launched in 1991, Italy's ETR in 1993. In Asia, South Korea began its rapid service in 2004, China in 2007. The latter country now has by far the longest network (42,000 kilometers by the end of 2022) of fast trains—including the longest direct link (2,198 kilometers), between Beijing and Guangzhou—and since 2017 the fastest average long-distance journey (Beijing–Shanghai, 1,300 kilometers), with the Fuxing train averaging 290 km/h and reaching maximum short-term speeds up to 350 km/h.[54] Moreover, in 2019 China announced the development of a new permanent magnet traction motor designed for speeds up to 400 km/h.[55] The US and Canada are the only affluent countries without rapid trains. The American Acela (Boston–Washington) can go up to 240 km/h, but its mean speed is a mere 116 km/h—no better than the best trains could manage a century ago.[56]

Speed gains in ocean shipping are best compared not in terms of brief maxima (be they under sail or under steam or diesel-powered) but as the time elapsed for specific crossings—and until the post-1970 rise of the Asian container trade to Europe and North America none of those was more important than the North Atlantic routes between the US (and Canada) and Europe. The baseline would be the commercial crossings by sailing ships: eastbound, three, sometimes four weeks; westbound (against the prevailing winds), usually six weeks. The first steam-powered eastward crossing was in 1833, when the Quebec-built paddlewheel *Royal William* made it from Pictou in Nova Scotia to London in 25 days.[57]

The first westward crossings, in April 1838, involved a race between two paddlewheels: *Great Western*, at that time the world's largest passenger ship, built by Isambard Kingdom Brunel (Britain's leading engineer) for the Great Western Steamship Company, and *Sirius*, a small vessel chartered by the British and American Steam Navigation Company. Although *Sirius* had a four-day head start, it barely beat (after averaging 14.87 km/h) the *Great Western*, taking 18 days, 14 hours, and 22 minutes, while the larger ship needed only 15 days and 12 hours.[58] Even these first crossings more than halved the sailing records, and steamships would keep posting faster times for the next 114 years.

A decade after the initial race, Cunard's paddlewheel *Europa* took 8 days and 23 hours in 1848 (averaging 21.8 km/h); by 1888 (with all ships equipped with propellers and higher-efficiency steam engines) the trip lasted just over 6 days; in 1908, the steam turbine–powered *Lusitania* won the Blue Riband (the unofficial award for the fastest crossing) with 4 days, 20 hours, and 22 minutes.[59] *Mauretania* claimed the title in 1909, and held it for 20 years before Germany's *Bremen* took it in 1929 with the first crossing averaging more than 50 km/h. *Queen Mary* posted the fastest pre–Second World War time in 1938 (57.4 km/h), and in 1952 the steam turbine–powered *United States*, the final record holder, made it in just 3 days, 10 hours, and 40 minutes (63.9 km/h). This means that typical westbound crossing times across the North Atlantic were cut from about 1,000 hours for sail ships to about 100 hours for steamships, a speed increase of an order of magnitude.

Jetliners (across the North Atlantic for the first time in 1958) rapidly ended the era of steam in intercontinental transport: by 1965, about 90 percent of all passengers were airborne; after 1,001 crossings, *Queen Mary* was retired in December 1967; and by 1968, 130 years after that memorable westward paddlewheel race, scheduled steamship travel was over and new ships were built for cruising.[60] Cruise ships, typically with a diesel-electric drive, have grown larger but not faster. In 2022, *Wonder of the Seas* set new records at nearly 237,000 tons, a maximum of 6,988 passengers and total power of 82 megawatts, but its cruising speed is 41 km/h.[61]

Bulk cargos (oil, coal, ores) aside, before the Second World War maritime shipping relied on general cargo vessels that were laboriously loaded with products packaged variously in boxes, crates, and sacks. Offloading was similarly slow, and the unprecedented level of intercontinental trade would have been impossible without a new, rationalized way of shipping. The use of steel containers (typically 20 feet or 6.1 meters long) began slowly during the late 1950s; the construction of vessels specifically designed for container transportation expanded during the 1960s; and the practice now dominates the business of long-distance ocean transportation, with vessels designed for speeds of 18–23 knots (33.3–42.6 km/h).[62]

Containers carry higher-value cargo including industrial products (machinery, tools), consumer items (electronics, apparel, kitchenware, toys), food (some in refrigerated units), and some raw materials. Heavy and voluminous cargo—fossil fuels (coal, crude oil and refined products, liquefied natural gas), ores (iron, aluminum, copper), cement, fertilizers, and unprocessed food (cereals, legumes, oilseeds)—are shipped by tankers and bulk carriers. These ships are now powered by large diesel engines (including the largest one in service with a power of about 80 megawatts) and their speed—13–19 knots for crude oil tankers, 16–20 knots for liquefied natural gas tankers, and 12–15 knots for massive bulk carriers—is typically a trade-off between two opposite demands: the speed to maximize the utilization of the vessels, and the number of vessels required to maintain a given service frequency.[63]

Container cargo (electronics, garments, food) has a much higher unit value than bulk cargo, and it should reach its markets as fast as possible (a shorter transit time also cuts inventory cost)—but, above a speed of 14 knots, the energy consumption (heavy fuel oil) goes up exponentially with sailing speed, and hence shippers and customers must figure out the most profitable fleet deployment (the number and type of ships assigned to specific routes) to maintain desired deliveries with the smallest number of vessels traveling at the most economic speeds: reduced sailing speeds may require an increase in deployed ships (or a decrease if larger vessels are substituted).[64] Moreover, speed optimization is one of the best options for reducing greenhouse gas emissions. Vessel speed optimization has thus become an important cost-reduction strategy in maritime shipping.

In the early 1970s, container vessels carrying fewer than 2,300 containers had speeds as high as 28 knots and consumed 300 tons of heavy fuel oil a day. Subsequent designed speeds of large container ships were 25–27 knots, but after the financial crisis of 2008 that dropped to 20–22 knots, and during the third decade of the 21st century some vessels have been operating at speeds up to 50 percent lower than their designed rate. Falling freight rates, rising fuel prices, and shipping overcapacity (due to weakened demand) explain this reversal, as operators must reduce sailing speeds to economically more acceptable levels. "Slow steaming" at speeds of 18–20 knots (33.3–37 km/h) has become more common. The energy (and emission) savings are obvious: above a speed of 14 knots, the energy consumed per unit of distance increases exponentially.[65]

An in-depth study of slow steaming by major Swedish shipping companies showed some surprising results: while expectedly the practice has resulted in 20–50 percent longer transit times, it has brought no increase in the reliability of deliveries and only marginal cost decreases.[66] While slow steaming results in indisputably lower greenhouse gas emissions, companies have not seen it as an explicit measure for better environmental performance, even as they have worried about their customers' willingness to pay the additional costs. And, as expected, companies dealing with made-to-order goods (apparel, toys) place a higher value on speed.

Given the mass and volume of maritime shipments (both in global aggregates and through individual ports), the speed of handling containers (bringing them to terminals, loading, unloading, and distributing them to intermediate or final recipients) is of the utmost importance. The loading and unloading of ships (the largest ones can now carry more than 20,000 20-foot, or 6.1-meter, containers) is done by large quayside cranes powered by large electric motors. Their lifting capacities range mostly from 40 to 80 tons, with maxima up to 120 tons.[67] With a single hoist/spreader they handle containers 20–45 feet long, and a dual hoist option lifts two larger or four smaller units. Overall loading/offloading speed is measured in moves: one move entails the time it takes to lift a container from the vessel onto the quay or vice versa.

But speedy offloading depends on adequate capacities of internal trucks to take the containers to a storage yard, as well as the availability of external trucks (or railcars) to take the containers out of a terminal. Obviously, these sequential and concurrent movements of large numbers of heavy units offer many opportunities for clever optimization but (as demonstrated in some terminals during the Covid years) everything can come to a halt as a weak link throttles the flow: scores of ships waiting to unload, striking terminal workers, a shortage of truckers, a lack of on-site container storage capacity—or disruptions caused by a pandemic. The backlog of ships waiting to unload at Los Angeles, America's largest container port, started in the fall of 2020 and it was not fully cleared until two years later.[68]

But overall, the speed, and the underlying logistics, of global cargo handling are admirable. Before the pandemic, about 11 gigatons of products were loaded annually around the world, with nearly 8 gigatons of dry cargo (dominated by iron ore, coal, grain, steel, and wood) and about 3 gigatons of crude oil, refined oil products, and liquefied natural gas.[69] The loading of both dry and liquid bulk cargos is now completely mechanized. Dry bulk cargos (such as coal and iron ore) are loaded by using a combination of rotary-wheel reclaimers, which pick the material from the stockpile and transfer it to a traveling and adjustable belt conveyor that discharges the cargo directly into a ship's hold (a slower option is to do it with cranes and grabs). Liquid cargos are pumped onboard, and depending on the pump and pipe capacities, even large oil tankers can be loaded in 24–36 hours.

As already described, the loading and offloading of container ships is a necessarily more complicated operation, but with new terminals, higher-capacity cranes, and better port logistics the world has been able to increase the annual shipments of containers (measured in the equivalents of 20-foot-long units) to nearly 70 million in 2000 and then to about 160 million in 2019, before the pandemic interrupted the growth—while worldwide container traffic rose from 225 million standard units in the year 2000 to just over 800 million in 2019.[70] This means that quayside cranes had to accomplish some 1.6 billion moves a year—or, somewhere around the world, day and night, 50 moves every second, some two-thirds of them in Asia (with 7 of the world's 10 busiest container ports being in China).

Road vehicles were historically the third category of transport whose speeds were multiplied by the adoption of fossil-fueled internal combustion engines. There were some early, unwieldy, and impractical designs of steam-powered vehicles, but the first road machines that used gasoline-fueled internal combustion engines were built only during the mid-1880s. Working independently and completing their first projects almost at the same time, Carl Benz, Gottlieb Daimler and Wilhelm Maybach filed their automotive patents on the same day, January 29, 1886.[71] Benz's three-wheel Patent-Motorwagen had a top speed of 16 km/h; Maybach's four-wheeler could reach 18 km/h.

Perhaps the best indication of how rapidly the subsequent designs developed is that Maybach's 1901 Mercedes (a four-door phaeton), introduced after Daimler's death in 1900 and considered the first truly modern internal-combustion vehicle, had a top speed of 72 km/h, and that contemporary cars designed specifically for racing, a newly popular spectator sport in Europe, could soon reach more than 100 km/h.[72] In 1899 the fastest racer (105.26 km/h) was Camille Jenatzy's *Le Jamais Content*, an electric vehicle powered by lead-acid batteries, but by 1902 the Mors Type Z, powered by a large gasoline engine, broke the record several times, eventually reaching 124.1 km/h, and in 1913 Perry Lambert, driving a Talbot design, was the first driver to top 100 mph (160 km/h).[73]

These speeds were limited to racing: at that time no country had extensive hard-topped roads and the existing surfaces could not sustain even modest speeds on longer intercity runs. But this infrastructural constraint began to ease in the US and Europe during the 1920s. By 1929, just before the great economic crisis began, the bestselling vehicles could do 105 km/h, and driving speeds of 70–80 km/h were common on newly paved intercity roads. Germany's *Autobahnen* of the 1930s were the world's first double-lane concrete highways. Before fuel shortages led to the imposition of an 80 km/h maximum in 1939 they had no maximum posted speed.[74] Construction of their American counterparts began only in 1956, and by 2022 the US Interstate system had nearly 80,000 kilometers of multilane highways—with maximum speed limits in rural areas mostly between 65 and 70 mph (104.6–112.6 km/h) and up to 80 mph (128.7 km/h) on specific segments in states including Texas, Oklahoma, Montana, and Wyoming.[75]

After the Second World War, Germany reverted to no-speed-limit driving on *Autobahnen*, and this usage has persisted through all subsequent oil price rises and economic downturns. But the practice has become greatly restricted, as many sections post 120 km/h limits, and 130 km/h is the recommended speed.[76] All post–Second World War waves of automobilization—an almost immediate resumption of the

US rise in car ownership, followed by European small car acquisitions in the 1950s and 1960s, the rise of Japanese automaking in the 1960s and 1970s and its eventual expansion into North American and global markets, and eventually the rapid rise of car ownership in post-2000 China—have been accompanied by lasting preoccupations with speed.

This has come in several forms. Even ordinary, mass-produced vehicles to be used overwhelmingly either in city driving (going from red light to red light at prescribed speeds, typically 60 km/h and maximum 80 km/h) or for intercity drives where the longest stretches are covered best by maintaining steady allowed speeds (mostly 60–120 km/h) or having cruise control do that, can go unnecessarily (and dangerously) fast. The Honda Civic, a bestselling small car whose different models I have been driving for decades, has speedometers going up to 260 km/h and its Type R can reach 272 km/h. Practically irrelevant record speeds achievable by production cars go far beyond that, with the Hennessey Venom F5 and two Bugattis (Bolide and Chiron) claiming speeds just above 300 mph.[77] And then there are the (for transportation, even more irrelevant) speeds of unique vehicles driven on salt flats: the latest record for a piston-engine vehicle (the Speed Demon, with a 3,156-horsepower Chevy engine) was set on August 14, 2020 at the Bonneville Salt Flats in Utah: the flying mile at 470.33 mph, and an exit speed of 481.576 mph.[78]

The other most encountered car-related speed metric is the time needed to accelerate from 0 to 60 mph (that is, to 97 km/h or 27 m/s) in America or (only slightly differently) from 0 to 100 km/h in the metric world. I have never clocked it myself, but my Honda Civic should do it in 5.8 seconds; the fastest acceleration for an internal combustion engine car belongs to Bugatti Chiron Super at 2.3 seconds; and powerful electric cars can do even better: the Pininfarina Battista, equipped with a 1.4-megawatt engine (a dozen times more powerful than my Honda), being the record holder with 1.79 seconds from 0 to 60 mph.[79] The obvious question is "who needs that, where and when?" (especially when purchasing that capability costs more than $2.2 million).

The third common manifestation of the speed-and-vehicle link is the lasting popularity of car racing in its various forms. Contrary to what most people might think, America's leading spectator sport is not football or baseball but NASCAR racing, with more than 75 million fans filling arenas capable of holding up to 190,000 people (it is also, with more than $3 billion in sponsorships, the most lucrative spectator sport).[80] Europe has its longtime favorites—the Monaco Grand Prix (since 1929) and 24 Hours of Le Mans (since 1923)—as well as many rally contests, and Formula 1 Grand Prix races take place around the world, including (in 2023) Azerbaijan, Bahrain, Brazil, Hungary, Qatar, and Singapore.[81] And, finally, there is the world of informal and illegal racing, common enough and sufficiently appealing to many viewers to generate movies, TV series, and podcasts.[82]

Better materials, better structural integrity, and mandatory seat belts (all based on Nils Bohlin's 1958 three-point design) and airbags (gradually introduced since the 1970s) have made all cars safer but cannot eliminate the fundamental link between speed and the risk of serious injury.[83] Four variables combine when stopping a vehicle: perception distance, reaction distance, brake lag distance, and braking distance.[84] Even for an alert driver, 0.75 seconds elapse between seeing and registering a hazard, and hence for a car traveling at 80 km/h even an instant subsequent stopping would take place after advancing 16 meters, or nearly four car lengths.

Reaction time (from brain to brake pedal) adds another 0.75 seconds and brake lag distance about 0.5 seconds, while the braking distance (from stepping on the brake to a full stop) depends on the vehicle's speed, mass, and length. At 80 km/h it is at least 64 meters, and considerably more on wet, snowy, or icy roads—but only 25 meters for a vehicle traveling at 50 km/h on a dry road. The combined reaction and braking distance increases from 14 meters at 30 km/h to 62 meters at 60 km/h and to 155 meters at 80 km/h, the latter distance virtually guaranteeing an accident.[85] Consequently, every mile per hour (1.6 km/h) of speed reduction is associated with a 6 percent decrease in traffic fatalities. I will return to speed and deaths in the final chapter.

Internal combustion engines have not only sped up all forms of transportation on land and water, but they also made flight possible.

In this case the baseline for speed comparisons is, obviously, zero. As we have seen, flapping flight is well beyond our anatomical structure and muscular capabilities, and gliders became practical only during the 1920s, after engine-driven airplanes had fought in the First World War and the first airlines had begun commercial service on intercity flights.[86] We could not fly before we deployed engines powering heavier-than-air planes—though, as already noted in the second chapter, some trained individuals eventually, starting in 1977, succeeded in making very light machines airborne by vigorous pedaling, and they were able to cover increasingly longer distances at low speed and close to the ground.[87] The successful development of flying machines came only at the beginning of the 20th century, but the progress, accelerated by two world wars, was rapid—and by the 1970s mass-scale air travel had emerged as one of the notable characteristics of modern globalized civilization.[88]

In 1903, an aircraft built by Orville and Wilbur Wright completed the first series of engine-powered flights. The Wright Flyer covered only short distances, with its longest low-altitude flight covering 260 meters in 59 seconds, a speed of just 4.4 m/s (less than 16 km/h).[89] Five years later, Glenn Curtiss won the Scientific American Trophy by completing the first public flight of more than 1 kilometer: he flew 1,550 meters in 1 minute and 42 seconds, averaging nearly 16 km/h—but by 1913 Gustav Hamel won the second Aerial Derby by flying 151 kilometers in 1 hour, 15 minutes, and 49 seconds, averaging 119.5 km/h, as fast aloft as the best locomotives (after nearly a century of development) could do at that time on rails.[90]

As the wooden planes quickly got not only faster but also slightly larger, they were turned into the first flying military machines of the First World War, and the world's first airlines had already been established before the conflict began. Right after the war, pioneering airlines were offering short flights with a few passengers in small and slow wooden airplanes. In 1920, the de Havilland DH.16, a four-passenger biplane used by the Dutch KLM (now the world's oldest operating airline), could reach 219 km/h.[91]

Longer connections came only after the introduction of all-metal fuselages and more powerful reciprocating aeroengines, but speeds remained limited. The British all-metal, four-engine Handley Page HP.42 airliner, introduced in 1930, could carry 24 passengers but its cruising speed was just 160 km/h and its range was only 800 kilometers.[92] When British Imperial Airways began to operate the London–Singapore link in 1934, its planes needed eight days and 22 layovers, including stops in Athens, Alexandria, Cairo, Baghdad, Basra, Sharjah, Jodhpur, Delhi, Calcutta, Akyab, Rangoon, and Bangkok.[93] Similarly, flying coast-to-coast in the US required three stops and more than 15 hours between New York and Los Angeles.

The first radical change came in 1936 with the introduction of the Douglas DC-3 (widely known as the Dakota), which could carry 21–32 passengers at a cruising speed of 333 km/h and had a range of just above 2,500 kilometers—not quite enough to make it from Chicago to San Francisco, but more than twice the distance between Chicago and New York.[94] Meanwhile, the first intercontinental flights were accomplished by *Graf Zeppelin*, the German airship powered by diesel engines. It made the first commercial passenger flight across the Atlantic on October 11, 1928, landing at Lakehurst, New Jersey, after 111 hours and 44 minutes, barely faster than *Bremen* (the Blue Riband record holder) the following year.[95]

In 1929, *Graf Zeppelin* circumnavigated the Earth, flying eastward from Lakehurst to Friedrichshafen in Germany and then on to Tokyo and Los Angeles, returning to New Jersey three weeks after its departure; in 1931 it started a regular passenger and mail service between Germany and Brazil; and in 1936 a larger airship, *Hindenburg*, began flying to the US: its catastrophic landing and fire in May 1937 ended not only that short-lived service but also the dreams of commercial airship transportation.[96] Even after putting aside the risks of flying with hydrogen-filled aircraft at altitudes that exposed the airship to potentially violent weather, it was clear that the maximum airship speed of 240 km/h was not the future of flight at a time when new airplane designs began to surpass 300 km/h and ranges had increased to thousands of kilometers.

Intercontinental commercial flights powered by reciprocating engines began just before the beginning of the Second World War with Boeing's 314 Clipper, a flying boat capable of carrying 68 passengers (or 36 in a sleeping bed configuration).[97] The flying boat was used (with stopovers) for a short-lived service between the US and UK, and for a Pan Am route (until 1941) between San Francisco and Hong Kong. The Clipper, powered by four air-cooled 14-cylinder radial piston engines, cruised at 303 km/h. Its range of 5,930 kilometers was more than enough for comfortable trans-Pcific hopping (San Francisco–Honolulu–Wake–Guam–Manila), but its ceiling (5,975 meters) was still in the lower troposphere with its changeable weather.

The first fighter planes powered by jet engines (gas turbines) appeared just before the end of the Second World War, and while they had no bearing on the conflict's outcome their development provided the foundation for postwar designs of both military and commercial jetliners flying close to, and above, the speed of sound.[98] Fast-flying speeds are usually designated in Mach numbers, the quotient of the aircraft's speed (true airspeed) and the local speed of sound. Recall that the latter speed is 340.3 m/s at sea level (at 15°C), but that it declines with altitude and with lower temperature: at 35,000 feet (10.7 kilometers)—the typical cruising altitude of modern commercial jetliners—the speed of sound is 295.4 m/s or 1,063 km/h.[99] As the speed of sound decreases with altitude, Mach numbers increase even as a plane's airspeed remains constant. Flying at Mach 1 is the transonic speed, everything below it is subsonic, everything above it up to Mach 5 is supersonic, and speeds above Mach 5 are hypersonic.

The speeds of wartime German (Messerschmitt) and British (Gloster Meteor) aircraft were close to 1,000 km/h, and that barrier was briefly broken (at 1,003.67 km/h) for the first time by the Messerschmitt Me 163A in October 1941. In July 1944, the Me 262 reached 998.5 km/h in horizontal flight; in November 1945, the Gloster Meteor flew at 975.87 km/h; and the sound barrier was broken for the first time on October 14, 1947, when Chuck Yeager piloted the Bell X-1 (released at 7.62 kilometers from the bomb bay of a B-29 plane over southern California) at a top speed of 1,126 km/h.[100] But the X-1 did not have a gas turbine; it was powered by a rocket engine burning nitrogen-pressurized ethyl alcohol and liquid oxygen.

The Korean War (1950–1954) was the first conflict in which both sides used fighter jets, but both the American F-86 Sabre and the Soviet MiG-15 could go supersonic only briefly in a dive: their designed maximum speeds were Mach 0.88.[101] The first American gas turbine–powered fighter plane capable of exceeding the speed of sound in level flight was the F-100 Super Sabre (flown between 1954 and 1971), designed to reach Mach 1.14.[102] This achievement was followed by periodic redesigns and upgrading, with most of the advances introduced in the US and Soviet Union/Russia. By 2020, only three countries were producing the most advanced fifth-generation fighter jets: America's Lockheed Martin F-22 and F-35, Russia's Sukhoi Su-57, and China's Chengdu J-20, with maximum speeds of Mach 1.6–2.0.[103] But these are not the fastest piloted flying machines. Even after leaving faster experimental planes aside, the fastest fighter plane in service in the early 2020s was the McDonnell Douglas F-15 Eagle, with a top speed of Mach 2.4 (the Soviet MiG-25 was faster but the old design stopped flying in 2022).[104]

The new era of jet (turbine-powered) commercial flight began with the British Comet, introduced in 1952. This (initially under-powered) turbojet could carry up to 44 passengers at a cruising speed of 740 km/h, but its design flaws led to a series of fatal accidents that ended its service in 1954.[105] When a redesigned version returned to

The McDonnell Douglas F-15 Eagle: no combat plane is faster.

service in 1958, it faced superior competition from the Boeing 707, America's first turbojet, which entered scheduled service in October 1958, flying from London to New York in less than 8 hours.[106] The plane's cruising speed was set at 890 km/h—and 65 years later speeds just below or barely above 900 km/h remain the norm even for the planes powered by the most advanced turbofan engines.

The first wide-body (twin-aisle) airplane, Boeing's famous 747 Jumbo Jet, had a cruising speed of 900 km/h; the latest Boeing (787, the Dreamliner) is rated at 903 km/h, as is the rather unsuccessful (too large) double-decker Airbus A380.[107] This means that the speeds of commercial flying have remained constant for nearly seven decades. No, this is not a display of ignorance: of course, everybody even slightly familiar with the history of flying remembers the Concorde, the British-French design that flew at supersonic speeds (maximum Mach 2.04, approximately 2,167 km/h) between 1976 and 2003—but that was never a commercial, revenue-earning plane, just a demonstrative prestige-garnering enterprise never flown for profit.[108]

Breaking the sound barrier has always been associated not only with creating a loud, offensive sonic boom on the ground (and hence excluding overland flight trajectories) but also with giving up any opportunity to have a profitable, long-lived commercial operation. Of course, airliners should fly as fast as possible to maximize their profit (that is, to minimize the cost per passenger-kilometer of travel). A new Boeing 787-8 costs nearly $250 million, and leasing it (at about $1 million per month) is more attractive to most airlines. In any case, these are large financial obligations for airlines to meet, and modern jetliners are scheduled to be in the air as much as possible and to fly as fast as is economical. Turnarounds, even for intercontinental flights, can be as short as two hours, and with two 10-hour legs a day a plane may be aloft for 20 hours—until the next periodic maintenance.

Greatly increased fuel costs at supersonic speed are the key factor limiting the speed of commercial aircraft. As already explained in the section on bird flight, the specific energy needs of flying are determined by finesse, the ratio of lift and drag, which is as high as 25 for a wandering albatross. As the speed of flying approaches the speed of sound, drag rises rapidly to more than 10 times its low-speed value.

The drastic drop in finesse at supersonic speeds on the one hand and the need to fly as fast as profitable on the other creates a very narrow performance constraint. Nobody has summarized this reality better than aeronautical engineer Henk Tennekes: "Mach 0.9 is an absolute maximum. These two constraints allow no latitude: a cruising speed of about Mach 0.85 is both the minimum and the maximum for long-distance airliners."[109] That is why the two latest, and very different designs—the Boeing 787 and Airbus A380—have identical design cruising speeds of Mach 0.85, with the Boeing 777 and Airbus A340 coming very close, respectively rated at Mach 0.84 and 0.83.

These constraints explain why the recurrent arguments in favor of airships or supersonic planes will remain limited to pop-science features and to failed startup companies. Airships are slower, and a lot trickier to operate, than high-speed trains with which they cannot compete on price: a rapid train can carry between nearly 500 (TGV) and more than 1,300 (Shinkansen) passengers. Similarly, no supersonic airplane could board nearly 300 passengers and, in addition, carry (as the Boeing 787-9 can) up to 60 tons of cargo (Concorde carried no cargo at all) and cross the Pacific at an affordable price. The speed constraint in long-distance commercial flying is clear: as fast as possible, but no faster (or slower) than economically unprofitable.

Of course, supersonic fighters can ignore this dictum as taxpayers (or rising deficits) make them possible at an increasingly steep price: the latest F-15EX costs more than $100 million, and although none of the F-15 series planes have been lost in actual combat, since 1975 more than 130 of them have been destroyed or damaged in serious accidents in the US alone.[110] But the costs have risen so much that some programs were ended early: only 195 Lockheed Martin F-22 Raptor stealth fighters were built between 1996 and 2011, and a 2017 report to Congress concluded that restarting production would cost more than $200 million per plane: the real cost might be on the order of a quarter billion dollars, an enormous price for a single Mach 1.8 plane.[111]

Any comprehensive review of the progress of machine-enabled speeds in agricultural and industrial production would easily fill a long book, but there is a way to cover these gains in a concise but convincing manner: by dispensing with a myriad of energetic, technical, and logistic specifics and concentrating first on rising productivities made (largely, or in large part) possible by the mass-scale adoption of engines and motors, and then on their combined effect in raising the rates of typical economic growth.

I explained in the preceding chapter how the gradual deployment of slow rotary machines—water mills and windmills—did, above all in some European countries, eventually reach the extent (that is, a higher unit capacity and a larger number of installations) that enabled the proto-industrialization of economies. But it was only when engines and motors almost completely took over tasks that were performed for millennia by animal and human muscles that the modern age truly began.

The adoption of tractors and combines accelerated a process that began with more powerful horses and steel machinery: internal combustion engines (mostly diesels after 1920) reduced the times needed for field tasks to small fractions of previously expected exertions. Plowing—energetically the most demanding field operation—was sped up by the use of more powerful tractors, able to pull wider plows, and harvesting was completely mechanized by combines offloading directly onto trucks.[112]

At the beginning of the 20th century, just before tractors and combines began to make the difference, wheat was produced on large Great Plains farms by using horses harnessed to steel machinery (plows, harrows, drills, harvesters) and threshed by steam-powered machines. The entire production sequence took slightly over 20 hours per hectare and it required 1–1.5 minutes of human labor per kilogram of grain; at the beginning of the 21st century, it took 2–3 hours per hectare and 2–3 seconds per kilogram of wheat.[113] This means that the time needed to produce a kilogram of wheat had been cut by at least 95 percent in a century—and the country's agricultural labor force declined almost exactly at the same speed: in 1900, 40 percent of Americans worked on farms; in 2000, only about 2 percent—a cut of 95 percent![114]

Agricultural labor liberated by the unprecedented speed of food production poured into cities, leading to record rates of urbanization and industrialization. And industries could accommodate this influx of labor, because the faster production of primary inputs (from fuels, metals, and sawn wood to woven fabrics, flour, and meat) made possible by the large-scale deployment of rotary machines could supply larger volumes of basic materials that could be turned into a wider variety of finished products, ranging from faucets and furniture to cars and clothes. There were three fundamental accelerators underlying the progress of all industrial sectors: faster drilling that enabled the exploitation of previously inaccessible oil and gas resources; the mass-scale electrification of many industrial processes; and the faster speed of machines making machines.

Recall that until the 1890s an ancient percussion process was the only way to drill exploratory and production wells. America's first rotary drilling rig was used in the Corsicana field in Texas in 1895, and the early designs relied on fishtail and circular-toothed bits that had good penetration rates only in softer rock.[115] The rotary cone drill, invented by Howard Hughes in 1908, had two conical-shaped rollers with extending chisel teeth that were able (to cite the patent granted in 1909) to "disintegrate or pulverize the material with which they come in contact and thus form a round hole in said material when the head of the drill revolves."[116] The two angled rollers rotated on their spindles while the entire bit rotated at the end of the drill string. Further improvement came in 1933, when Floyd L. Scott and Lewis E. Garfield patented cutters with three rotating drill cone bits, further speeding up the drilling while also providing better support on the well bottom with reduced vibration.[117]

While percussion drilling could progress, in softer rock, no more than 1.5–3 meters a day, modern rotary cone drills can average 50–150 m/day, a 15- to 100-fold improvement.[118] Moreover, strong drilling rigs can support the heavy drilling strings (sections of connected steel pipes) needed for deep wells, and powerful engines (usually diesels) can rotate those drilling strings at 50–150 rpm to complete, routinely, wells between 2–4 kilometers deep even in hard rock—an unthinkable feat with percussion drills. Rotary motion used to produce liquid and gaseous fuels makes it possible to use these fuels (after suitable processing and refining) to power gasoline- and diesel-fueled

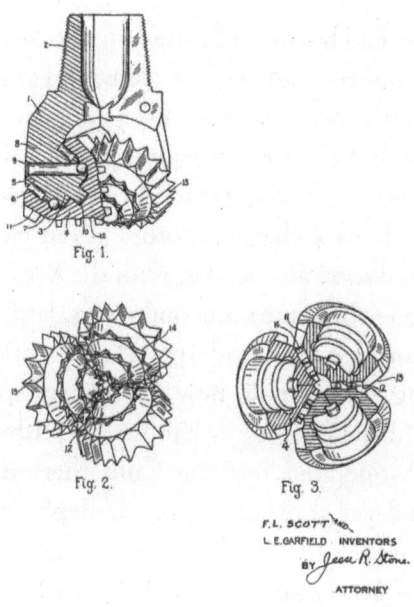

Dec. 4, 1934.   F. L. SCOTT ET AL   1,983,316

THREE-CONE BIT

Filed April 17, 1933

Fig. 1.

Fig. 2.   Fig. 3.

F.L. SCOTT AND

L. E.GARFIELD INVENTORS

BY *Jesse R. Stone.*

ATTORNEY

Faster drilling: drawings of the rotary tricone bit patented in 1934.

reciprocating engines as well as the most powerful modern rotary engines, stationary gas turbines, and jet engines.

The second fundamental accelerator of industrial growth was the replacement of animate power and steam engines by electric motors. This fundamental advance came with multiple benefits. Electric motors are the very pinnacle of rotary power, as they combine high efficiencies (often more than 90 percent for large units) with reliability, long service, and outstanding flexibility, even as they deliver a desired power ranging from fractions of a watt to tens of megawatts at speeds (as already noted) fit for an enormous array of final uses.

In American manufacturing, this transformation took place in just three decades: between 1899 and 1929, the sector's total mechanical power quadrupled, the capacity of electric motors increased almost 60-fold, and their share rose from just 5 percent to 82 percent of the total installed capacity.[119] This surge of electrification nearly doubled American manufacturing productivity, and as electrification grew more pervasive (in terms of total motors rather than total capacity installed), another productivity doubling took place by the late 1960s.[120]

Electric motors had begun to be used on railroads in the US and in some European countries during the 1890s, and after the First World War electrified railways came to Australia, some Asian countries, and South Africa, but their progress was delayed by the adoption of inexpensive and powerful diesel traction. As described earlier in this chapter, starting in 1964 electric motors began to speed up railway transportation, and soon afterwards, with the rise of a new electronic economy based on solid-state semiconductors, large numbers of electric motors began to power widespread automation, ranging from endlessly running escalators in new subway systems and photocell-activated doors (dispensing with handles or pushing) to coffee and hot water touch-operated machines and thermostat-regulated air conditioners. And yet another mass-scale deployment of tiny electric motors was needed to facilitate instant communication: no other class of motors is now more widespread than the hundreds of millions of the thinnest (less than 2-millimeter) micromotors used to activate vibrators in mobile phones.

The third fundamental accelerator of industrial production was the greater speed of machining. Traditional materials used for cutting and turning could work well only with wood and softer metals, but once advances in metallurgy made it possible to produce inexpensive steel from molten pig iron (first by the Bessemer process, and later in open-hearth furnaces), it was essential to come up with materials that could cut the metal at increasing speeds. And this need became even more important with the introduction of stainless steel (alloyed with chromium, first by Harry Brearley in 1912), steel alloys containing nickel and cobalt, and with the use of titanium and tungsten for the most demanding applications, ranging from surgical instruments to aircraft parts.[121]

Many products are made by casting hot materials into forms, by forming—(pressing and extruding), and most recently by additive manufacturing (the computer-guided layer-by-layer deposition of hard materials)—but cutting remains the foundation of modern metalworking, and the large global machine-tool market is still dominated by metal cutting. Meeting modern cutting needs is complicated by the fact that a material's hardness (wear resistance) is inversely related to its toughness (chipping resistance): high-speed steel, the least expensive choice for cutting tools, has a high chipping resistance and low wear resistance; ceramic materials wear much better but are easier to chip.[122]

The hardest, high-resistance alloys are also difficult to work with because their machining generates heat that cannot be—due to their low thermal conductivity—readily transferred either into the workpiece or into generated chips, posing the dangers of overheating. Despite these complications, high-speed machining (with conventional speeds exceeded 5- to 10-fold) became a commercial reality by the end of the 20th century, with spindle speeds of up to 12,000 rpm and a maximum cutting rate around 2,000 m/min. The cutting of new composite materials (including carbon fibers) is easier, but it brings the risks of fiber pullout, delamination, and high surface roughness.[123]

Konrad Wegener and his group at Zurich ETH summarized the history of metal-cutting advances by plotting the successive performances of the best cutting materials when machining a common metal: general-purpose medium carbon steel (CK45), which is sold in round bars or plates and is used for making strong, wear-resistant cutting tools.[124] In 1870, carbon cold-work steel would cut that alloy at 5 m/min; by 1910, high-speed cutting steel would run at 25 m/min; and by 1940, sintered carbides could cut at 100 m/min. By 1970, sintered cubic boron nitride was available: boron and nitrogen are pressed, with the help of binding materials, under extremely high temperatures and pressures to form the world's second-hardest substance (diamond remains the hardest), which can cut at close to 2,000 m/min. This means that the highest cutting speeds in metal machining rose about 400 times after 1870, and at least 200 times during the 20th century.

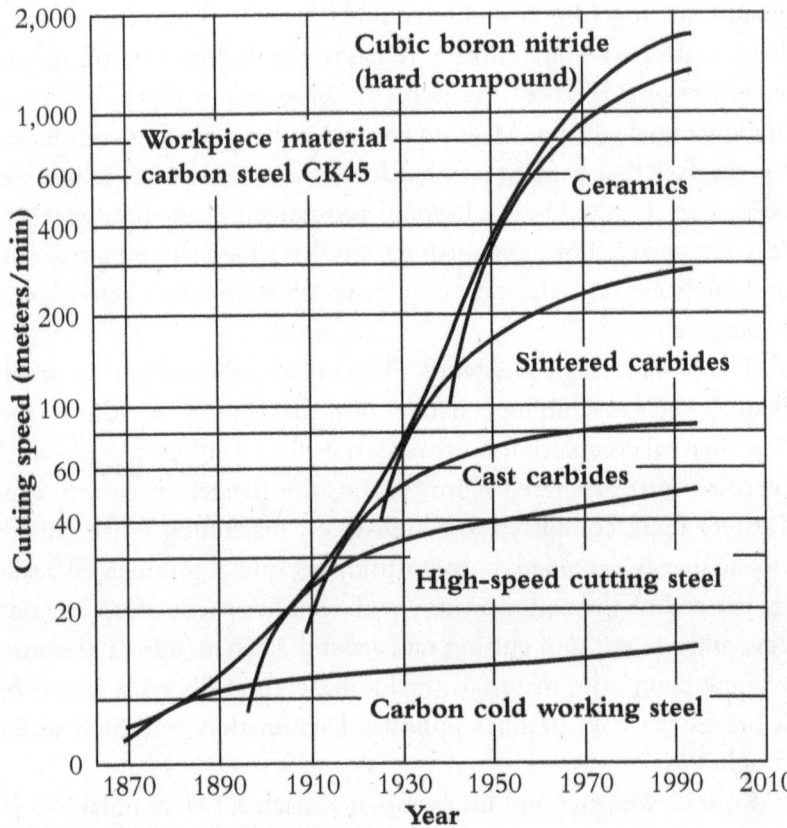

More than a century of higher machining speeds: from carbon steel to ceramics, from less than 10 to more than 1,000 meters cut per minute.

   This recounting of fundamental modern speed-ups of basic machines, processes, and techniques gives us a necessary and solidly quantitative foundation for the concluding evaluations to be presented in the final chapter. But first, two detours. First (and briefly), I must note one hidden innovation that has made possible many of the accelerations brought about by modern machines, and then (in more detail) I must quantify the most impressive of all modern (*sensu lato*) speed (performance) increases: the truly astounding transformation of long-distance communication and the mass-scale processing, dissemination, and storage of data, with the latter term now subsuming everything from texts and measurements to statistics and still and moving images.

This hidden innovation embraces a large class of mechanical bearings, now a universally deployed means of reducing friction due to the advantages of rotary (rolling) over linear (sliding) motion. By interposing bearings (rolling elements) between sliding surfaces, resistance is reduced, but the arrangement must reckon with high stresses in the restricted surfaces mediating the load transmission: that is why the use of ancient wooden ball bearings, best documented in a sunken Roman ship, was limited, and why conceptual bearing designs by Leonardo da Vinci and Galileo Galilei were not turned into practical devices.[125] In 1740, John Harrison invented tiny, caged roller bearings for his H3 chronometer; in 1794, Philip Vaughan, a Welsh ironmaster, patented the first iron ball bearing placed between a carriage wheel and its axle; and in 1862 came the invention of precision ball bearings for bicycles that reduced the friction to as little as a tenth of the previous value.[126]

Iron was inadequate to bear the enormous contact stresses (1.4 giganewtons per square meter) of high-duty bearings, and the need for special steels led to many metallurgical and manufacturing innovations pioneered by new precision ball bearing companies—most notably the Hoffmann Manufacturing Company and the Timken Roller Bearing Axle Company (both established in 1898) and Svenska Kullagerfabriken (SKF) in Gothenburg, the world's largest bearing maker (established in 1907).[127] Since that time, tens of billions of ball and roller bearings have become hidden but irreplaceable enablers of rotary speed. Most people will think immediately about the ball bearings in cars—in the engine, transmission, gearbox, wheels (turning at about 800 rpm in a vehicle doing 100 km/h)—but they are found in all transportation machinery, from unicycles and tricycles to jetliners and off-road trucks.

They are in just about every appliance, large and small (refrigerators, washers, dryers, vacuum cleaners, food processors, coffee grinders), in every electrical tool (from construction equipment to dental drills, where they sustain speeds of up to 400,000 rpm), in space (weather and spy satellites), and on Mars (in robotic rovers). And if much of modern speed depends on bearings, all unsealed bearings depend on lubricants, mineral or plant oils, or greases (oils with thickeners). Special greases are needed not only for high-speed (10,000–50,000-rpm) motors; a large new market has been created by wind-powered electricity generation. Wind turbines turn slowly (15–25 rpm) but their bearings are subjected to large stresses and must cope with large temperature changes. In the 21st century, speed (any speed) and grease are thus as closely connected as they were in the era of horse-drawn stagecoaches, wagons, and carts.

An overwhelming speed surge: communication, data, calculations

Information flows are now accomplished by encoding strings of symbols that are processed, exchanged, recorded (on magnetic, optical, or mechanical media), and dispatched to final users at speeds equaling or nearing the speed of light—hence essentially eliminating distance as a factor in human communication. The first practical steps toward these remarkable accomplishments began with the first commercial telegraph links almost exactly a century before the first (secret, voluminous, and still rather slow) electronic computers were deployed in the US and in the UK during the Second World War, and while post-1950 advances had their roots in the theoretical work carried out by James Clerk Maxwell during the 1860s and were possible thanks to the subsequent efforts of generations of physicists, the speed of the introduction and global adoption of these innovations has been truly unprecedented.

But to appreciate information speed gains we must start with speech, one of the three most consequential human attributes—together with upright posture and extraordinarily large brains. Spoken English has a frequency of 150–190 words per minute—that is, a speed of 2.5–3.2 words per second.[128] As with any other form of information transfer, this speed can also be expressed—in the most fundamental manner—in bits and bytes per second. A bit is a unit storing just a 0 or 1. Eight bits (b) make one byte (B), able to store one typed character. There are significant differences in syllabic information density between languages: Japanese has only about 5 bits/

syllable, English just over 7 bits/syllable (nearly 1 byte), and six-tonal Vietnamese has 8 bits/syllable.

There is also a noticeable range in syllabic pronunciation frequency: Italians can manage up to 9 syllables per second, Germans only 5–6. But despite these differences, speakers in 17 tested languages transmitted information at about the same rate.[129] Obviously, some languages sound noticeably faster than others, but when—in order to get the total information flow per second—the spoken rate is multiplied by the bit rate, the results are astonishingly consistent, tending toward the same speech speed of 39 bits (39.15 bits, to be exact) or about 5 bytes per second (Bps). The finding is much less surprising when we consider the identical anatomy of the human speech tract—a series of joined cavities (throat, mouth, nose) and vocal cords, nostrils, and lips.

Writing is the oldest way of physically transferring verbal information, and its speed ranges widely from the first attempts by preschool children to the routine pace of skilled copyists. English words average 5.1 characters (means are slightly higher in French and Spanish, and significantly higher in German), and writing 24 words per minute will be transferring information at 16 bits per second (bps); skilled writers (not stenographers) could go twice as fast, slow writers would proceed at less than a third of that rate, and the monks who created the elaborate majuscule scripts of the early Middle Ages were slower still.[130]

The short-term (less than a minute) information throughputs of human speech (39 bps) and rapid writing (32 bps) are higher than the information rates of other mental exertions: those range from about 5 bps for binary digit memorization and choice-reaction experiments (with participants reacting to different stimuli by selecting the appropriate response) to as much as 30–50 bps for object recognition. Consequently, when unimpeded by any constraints, the speed of human thinking typically amounts to no more than 20 bits of information over a few seconds, averaging 10 bps or less: when compared to the rates at which our peripheral nervous system is capable of absorbing information from the environment (on the order of gigabits/s) we must conclude that life at 10 bps truly means "the unbearable slowness of being."[131] Moreover, this enormous (about 100 million times) difference between information absorption and information processing remains largely unexplained.

Printing (since the late 9th century CE in China and since the mid-15th century in Europe) made a fundamental difference, but it took a while to get to very high speeds of information transfer. When printing was limited to pulling off individual pages from inked woodblocks (also as long as paper production was done laboriously, sheet by sheet), its rate of reproducing information could not surpass a few scores of sheets per hour for single-color prints. Mechanical printing machines (first by Gutenberg in 1455) sped up the process to 240 sheets printed per hour, but rapid paper-making came only with Fourdrinier's invention of a continuous machine, and rapid printing only with the first steam-powered printing press in 1814: it could run off 1,000 impressions per hour, by 1849 the new rotary presses could print 20,000 sheets per hour, and by 1902, when Robert Hoe reviewed the history of printing presses, his largest machine could print 96,000 4-, 6-, or 8-page papers and 72,000 10-page papers per hour.[132]

Reasonable speed multipliers are thus 10-fold (an order of magnitude) when going from woodblocks to the first mechanical presses, 100-fold when going from those presses to the first generation of the mid-19th-century rotary machines, and a further order of magnitude between 1850 and the first decade of the 20th century. Data transfer velocity estimates (in Bps) depend on several concatenated assumptions (size of the printed sheets, number of words per page, the actual speed of pulling off or printing, later also of the folding and binding of sheets), but for the first rotaries the rate (assuming 3,000 words or about 15,000 characters, or 15 kilobytes, per sheet) was about 5 megabytes per second, rising to more than 50 MBps before the First World War.

After a lengthy period of experimentation and short-distance trials, the first commercially viable electrical telegraph began to operate between Washington, DC and Baltimore in May 1844; the first, failed, transatlantic link came in 1858, and the first successful one (in 1866) was preceded by the first transcontinental telegraph line

(Western Union in the US) in 1861.[133] Just 15 years later, Alexander Graham Bell was granted his first telephone patent, and by 1887 Heinrich Hertz had generated electromagnetic waves by a large induction coil that received a pulsed voltage from batteries.[134] Before the First World War, the first broadcasts crossed the Atlantic; after the war, radio ownership in the US and Europe rose rapidly (by 1930, nearly half of all American households had a set), and the first experimental TV broadcasts took place during the 1930s.[135]

The Second World War saw the first (secret, massive, and comparatively slow) electronic computers, which relied on large numbers of vacuum tubes, but by the mid-1950s solid-state electronics began making commercial inroads, the first integrated circuits were patented, and in 1965 (as the first commercial satellites were transforming global telephony) Gordon Moore forecast the doubling rate for the number of components on an integrated circuit.[136] Intel's first microprocessor was released in 1971, and the 1970s saw the growth in size and speed of large supercomputers as well as the advent of small personal machines.[137] These trends intensified during the 1980s, the decade of the first portable telephones; email, the Internet, and search engines marked the 1990s; and mobile telephones began their rapid diffusion with 3G models after 2001.[138]

The history of these advances that created a new world of electrical and electronic communication and information flows (texts, data, images) has been reviewed many times, both in detail and in many shorter overviews, and here I will recount only its changing speeds. The plural is apposite not only because of the range of speeds to be quantified (from supercomputer calculations to global data transfers and from downloads to personal computers to uploads of images) but also because we must consider two different rates: physical speed (distance/time) and various rates of *sensu lato* speeds—that is, tasks (computations of units of information transmitted) accomplished per unit of time.

People can always talk intelligibly across scores of meters (and for more than 100 meters under the ideal circumstances) or shout (or whistle) and be heard far further, but the speed of sound, 343 m/s at 20°C at sea level, declines slowly and linearly with altitude.[139] Once we began to carry information by electricity transmitted through wires (telegraphs, landlines, undersea cables) or by electromagnetic waves broadcast through the air (radio, TV) or sent via special cables (coaxial, optical), it could travel at speeds very close to the speed of light—that is, 299,792,458 m/s, or about 0.3 meters per nanosecond (a billionth of a second).

But that maximum velocity is possible only in a vacuum, and the velocity of propagation (VOP, or velocity factor) for different cables is determined almost completely by the insulation materials (dielectrics, commonly different types of plastics): it is similar for copper and glass, with the VOP being at least two-thirds of the speed of light, or some 200,000 km/s; the VOP for solid Teflon is about 0.7 and for a closed-cell foam dielectric it is 0.9.[140] Coaxial cables are used to carry telephone, TV, and Internet traffic, and by 2020 they had reached the homes of about 85 percent of the US population. They are made up of a copper core, insulation, a copper shield, and an outer plastic layer. Fiber-optic cables (reaching 43 percent of US households by 2022) are more expensive to install and are made up of plastic or glass; they transmit modulated light across very long distances and have superior data transfer velocity.[141]

But before I review the latest electronic speeds, I should quantify comparable pre-electric and early electric baselines for transferring information across long distances. The only pre-electric choice to do that between continents was the physical transfer of printed or written texts by sailing. Imagine young Benjamin Franklin, a printer devoted to books, sailing back from England to Philadelphia in 1726 and carrying a complete version of Shakespeare's plays and sonnets (about 5.3 megabytes).[142] That trip took 83 days or about 7.17 million seconds, resulting in a physical displacement speed (spatial transfer rate) of about 0.75 Bps. And even if he were to pay for shipping 100 similarly sized tomes comprising the essential knowledge of the age, the rate would have been just 75 Bps: to transfer that amount of data (530 megabytes) to my computer in the center of Canada from London would take (at the rate of about 50 Mbps) less than a minute.[143]

And the first successful form of long-distance signaling (but only across hundreds of kilometers) was even slower than shipping tomes on a sailing ship. For millennia, nearly all long-distance communication (the limited use of simple smoke signals or drumbeats being the only exception) could not be separated from travel—that is, from walking, running, or (as we have seen) often elaborate and efficient messenger and postal services on horseback and (as just illustrated) by sailing. The first indirect transmission of information came surprisingly late, on March 2, 1791, when Claude and Ignace Chappe demonstrated Claude's optical telegraph by sending a brief message ("If you succeed, you will soon bask in glory") across 14 kilometers in 9 minutes.[144]

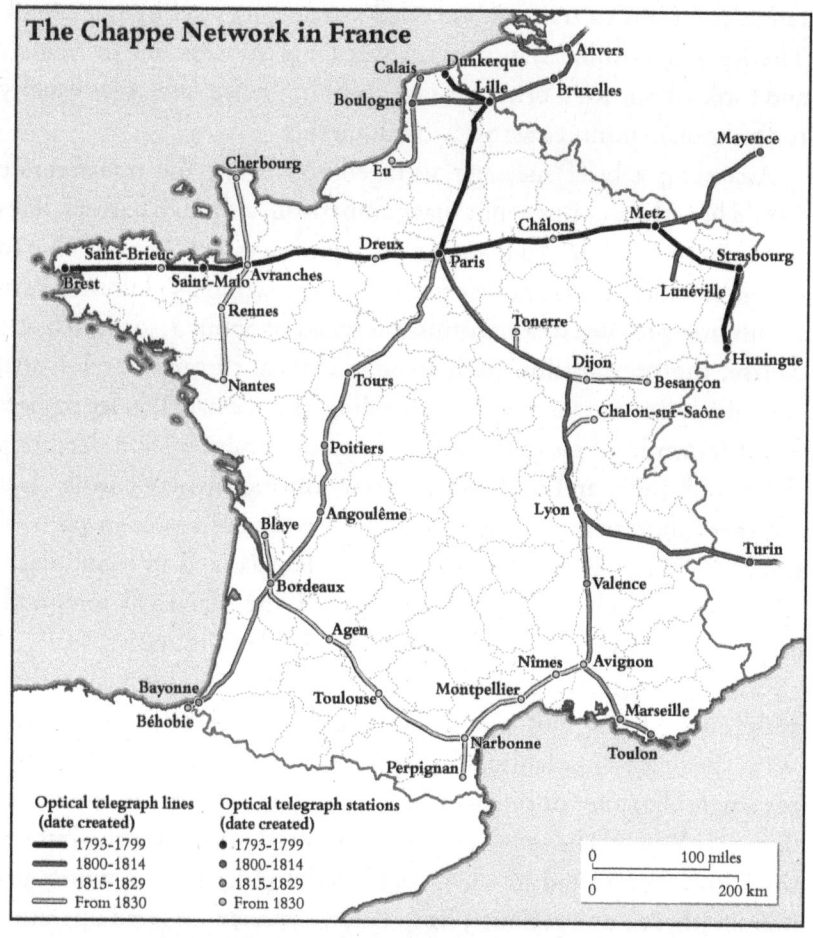

The national network of Chappe's optical telegraph.

The French text of nine words—*si vous réussissez vous serez bientôt couvert de gloire*—has 47 characters, and sending 47 bytes in 9 minutes amounts to less than 0.1 Bps. The first (Paris–Lille) optical telegraph line was ordered in August 1793 by the infamous Committee of Public Safety just a few months after its establishment, and the line was first tested on April 30, 1794, a few months before Robespierre's execution in July 1794. Eventually, the French government built a national system of 534 towers (spaced 5–20 kilometers apart), spanning nearly 5,000 kilometers and linking the capital to the Mediterranean, the Spanish border, Brittany, Alsace, and Belgium. The stone towers had tall masts topped with a cross beam and two arms that were set in different positions to transmit encoded signals observed by telescopes. The longest possible transmission was from Amsterdam to Venice, and took 1 hour for a brief message; within France, messages usually took about 10 minutes across 250 kilometers.

Assuming a brief message averaged 200 bytes, the transfer rate would be only 0.33 Bps—but that comparison, although correct, does not capture the radical advantage of the new system that was able to bridge distances at an unprecedented rate: bridging 250 kilometers in 10 minutes prorates to a transmission speed of about 1,500 km/h. Of course, the transmission itself (seeing a signal through a telescope) was nearly instantaneous, but encoding the messages (telegraphers could transmit no more than three symbols a minute) and decoding them took time, and it all worked only during the day and in clear weather. Shorter optical telegraph links were built later in parts of England, Spain, Germany, Sweden, and Russia, and in some places the system survived until the advent of the electrical telegraph (some stone towers in France are still accessible to tourists).

Similarly, the telegraph could not transfer words faster than people could talk, but mere talking could never bridge an ocean. With the first transatlantic telegraph cable of 1858, it took 2 minutes per single character or one word every 10 minutes—that is, no better than 0.33 Bps.[145] The second cable was severed during laying, but the third link, completed in 1866, succeeded and handled 8 words per minute (that is, up to about 3 Bps). Later, experienced hand operators could transmit 25–40 words per minute (up to 10 Bps), and automatic telegraphs were invented to speed up the process to 60–120 words for the ink recording version and as many as 1,000 words by Thomas Edison's invention that combined a perforator (about 35 words a minute) and a high-speed transmitter (up to 1,000 words per minute) at the transmitting point, and a metal stylus marking treated paper and leaving behind dots and dashes.[146]

Transmitting 1,000 words per minute raised the rate to above 200 Bps, but for shorter messages the slower manual option remained faster than the preparation of perforated tape and the transcription of the code. In the end, the telegraph enjoyed unchallenged dominance for only about a decade: as already noted, in 1876 (again, as in the telegraph's case, after decades of failed pioneering attempts going back to the 1840s) Alexander Graham Bell received the first telephone patent. Obviously, this way of direct spoken communication could not be generated faster than talking, but the telephone was destined to surpass the telegraph because it required no coding and decoding and offered flexibility, immediacy, and privacy at near-light speed.

Again, we must consider the speed of transmission in terms of both the transfer rate and the distance bridged by the communication. The transfer rate of instantaneously relayed telephone conversation was at 5 Bps, but the distances across which it was (almost instantaneously) transmitted increased from about 300 kilometers in 1884 (Boston to New York) to nearly 2,000 kilometers by 1898 (New York to Kansas City); the first transcontinental call (more than 4,000 kilometers) became possible in 1914; and telecommunications satellites (starting with Telstar in 1962) eventually made affordable automatic intercontinental calls a routine experience. At the same time, during the late 1970s, affluent countries began entering a new era of mass-scale computing whose advances eventually tied all information flows into a new electronic web dominated by rapid data transfers, enormous data storages, and an increasingly unmanageable intensity of data flows.[147]

The early advances of the telephone era coincided with the introduction and use of the first calculating electromechanical machines, the Second World War brought the first electronic computers, and simple transistors were followed by integrated circuits and by microprocessors—and the crowding of components on microchips

proceeded for decades at high exponential rates. Because of the range
of final uses and because of rapid qualitative changes (from vacuum
tubes to simple semiconductors to highly integrated circuits),
measuring computing advances across different techniques is not a
straightforward matter.

William Nordhaus defined his use of computer power "as the
number of times that a given bundle of computations can be performed in a given time."[148] Using this metric, we go from 1 unit per
second for mental calculations to about 6 units/s for the mechanical
calculators of the late 19th century to nearly 20 units/s for pre–
Second World War electromechanical devices.

The first generation of electronic computers (until 1949) raised
performance to about 1,700 units/s, two decades later it soared to
nearly 3 million, and by the end of the 20th century it stood at 40 billion and soon afterwards rose to trillions: this implies a performance

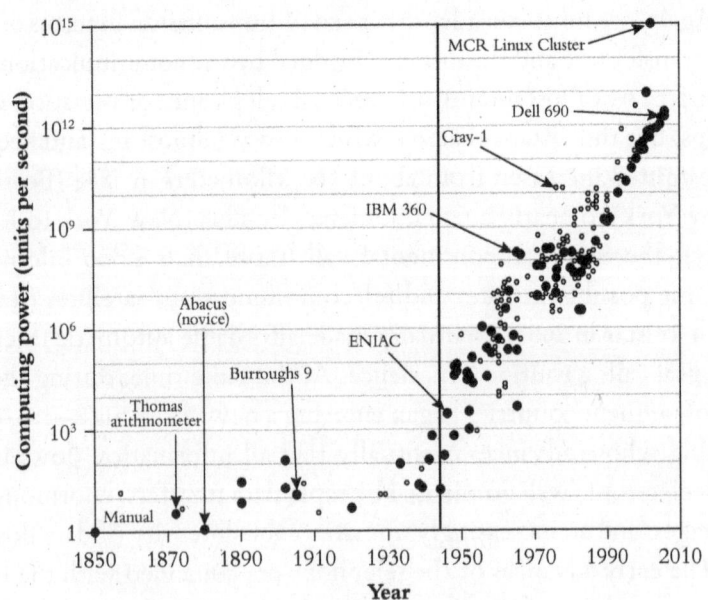

**The progress of computing power measured in
computations per second (CPS)**

From one calculation per second (mental/manual) to trillions of calculations
per second 120 years later.

rise of 12 orders of magnitude (tens of trillion times) in less than 120 years.

The processing speed of large mainframe computers and super-computers (starting with the CDC 6600 in 1964) is best expressed in terms of floating-point operations per second (flops): these are any mathematical operations (addition, subtraction, etc.) involving numbers with decimal points that take longer to execute than those involving simple binary integers.

The measure was introduced by Frank H. McMahon at Lawrence Livermore National Laboratory to compare the progress of supercom-puter capabilities, but it is revealing to view these capabilities against three baselines: ordinary calculations, the first large US machine, and the first commercial computer. Depending on their complexity, head-and-paper calculations proceed at merely 0.01–1 flop; in 1945 ENIAC processed about 500 flops, while the 1951 UNIVAC man-aged about 1,900 flops (nearly 2 kiloflops). One megaflop (1 million flops) was reached a decade later, and then came the still-advancing array of supercomputers: from the 160 megaflops of Cray-1 (1976) to IBM's ASCI White (2000) with 7.2 teraflops, and to El Capitan at the Lawrence Livermore National Laboratory which reached 1.742 exaflops in 2024.

The speed trajectory from 1 megaflop in the early 1960s to 1.7 exaflops in 2024 implies an increase of 12 orders of magnitude, or

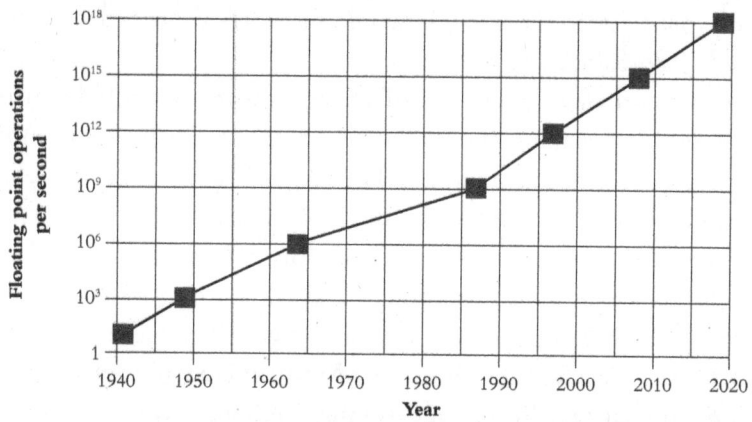

Computing flops.

a trillion-fold in six decades.[149] This has been the greatest speeding-up in history: I am not aware of any other increase in performance that could compare with this spectacular (and still-continuing) rise. Of course, supercomputers (now costing more than $100 million for the latest Hewlett-Packard designs) are beyond the reach of common users, but the speed of ordinary personal computers, although obviously no match for these record-breaking machines, has also increased by many orders of magnitude.

Its usual measure has been the number of cycles executed by the central processing unit (CPU) every second. One cycle per second is 1 hertz (the unit named after the pioneer of electromagnetic wave research) and hence the clock speed is measured in multiples of that basic unit. While it is not an automatic indicator of the superior performance of a particular computer (higher-speed designs can be outperformed by lower-speed machines that handle the instructions more efficiently), both across decades and within the same CPU brand and generation the higher clock speed means a faster over-all performance.[150] Clock speed rose from 100 hertz (0.1 kilohertz) in the ENIAC (1945) to 2 megahertz for the transistorized designs of the late 1950s; IBM's first personal computer (1981) had a clock speed of 4.77 megahertz, speeds above 1 gigahertz were reached by the beginning of the 21st century, and by 2020 the dominant speed in personal computers was 4.2 gigahertz. Going from 1945 to 2020 implies a speed gain of seven orders of magnitude (42 million times); from 1981 to 2020 the increase would be nearly three orders of magnitude (880-fold, to be exact).

And any review of modern electronic advances would be incomplete without tracing the evolution of speeds with which mobile phones transfer information.[151] For the first generation of cellular phones—heavy, long-charging devices with very limited reach—it was at the rate of 2,400 bps (2.4 kbps); for 2G (starting in 1991) it was nearly 100 times faster at 0.2 Mbps, and 3G (starting with the service provided by NTT DoCoMo in Japan in 2001, the first one capable of surfing the Net) had speeds up to 2 Mbps. 4G (2009) brought smartphones (up to 12.5 Mbps), and 5G (first in South Korea in 2019) was more than an order of magnitude faster (169.4 Mbps). 5G also has the shortest latency, which is the time span needed for a phone to send a message and get a response: just 10 milliseconds, compared to 50 milliseconds for 4G, undoubtedly a feature that has contributed to the global addiction to these fast mini-machines.[152]

That these technical accelerations created a new world is obvious, but some of the claims based on these realities should be questioned. Some commentators have elevated speed to the reigning preoccupation of the modern era, and see it as the very zeitgeist that changed human relationships and visions and led to time-space compression.[153] Of course, neither time nor space can be compressed—only the speed of specific developments can increase and so can their frequency, allowing us to reach more distant destinations and to experience more stimuli and more events in the same period of time. In the next chapter I will take a critical look at these speed-related claims and the realities.

But before ending this chapter I will, as I did in the previous chapter for premodern times, offer a brief review of the enormous increases of the instant—or near-instant—rate of modern killing. Here is the speed increase (in rounds fired per minute) for individually operated weapons: 2–3 for infantry rifles used during the Napoleonic Wars; 15 for Lee–Enfield rifles during the First World War (when Vickers machine guns could fire 600 rounds but needed six men to operate them); and 1,800 for the MG-42, the German light machine gun of the Second World War that was often operated by a single soldier.[154] This means that between Waterloo and Stalingrad, the capacity of an individual infantry soldier to kill and maim rose a thousandfold (three orders of magnitude).

But these potential killing speeds have been vastly surpassed by bombs delivered by airplanes and, since the late 1950s, put into the tips of (short-range to intercontinental) missiles. We will never know the exact casualty totals, but the firebombing of Tokyo (during the night of March 9–10, 1945) killed, within hours, at least 90,000 people, while on the days of nuclear bombing at least 70,000 people died in Hiroshima (August 6, 1945) and 25,000 people in Nagasaki (August 9, 1945), followed (in both cases) by large subsequent fatalities.[155]

And simulations indicate that about 34 million people would die, and about 57 million would be injured, within the first few hours of a nuclear exchange between Russia and the United States.[156] That means that maximum one-day wartime deaths have risen from about 10,000 at Waterloo to the potential for tens of millions of casualties in a nuclear exchange: again a thousandfold increase—and a reminder of the continuing deployment of human ingenuity to the end of faster ways of killing.

5. Parting Musings

Obverses, limits, aspirations

Every desirable accomplishment has its downsides, and when taking speed in *sensu stricto*, no other speed-related loss—of life, health, property—is now globally more common than the price paid for road accidents. Would we tolerate the continued use of a prescription drug whose side effects killed, year after year, more than a million people? Would we live, decade after decade, with a building design that led to recurrent structural collapses killing that many people and destroying their homes? Moreover, the quest for higher road speeds has led to the large-scale physical reordering of cities to accommodate fast vehicle traffic by ramming wider roads through neighborhoods and by building sometimes veritable mazes of freeways and dizzyingly complicated (including multilayer) interchanges. The resulting environmental impacts and lower quality of life (above all, air and noise pollution) have been all too evident.

Another notable downside caused by higher L/T rates is due to the higher speed of conveyor belts used in facilities ranging from those warehousing, sorting, and shipping goods to large-capacity meat-packing facilities.[1] In the latter case, the high rates of injuries (and even deaths) are due to the higher speed of killing animals and dismembering their carcasses by workers standing along moving conveyor belts, in plants now common in all modern carnivorous societies. In the former instance, those rapid and repetitive tasks lead both to physical injuries and to mental stress. Amazon is the best known and (with its rate of serious injuries running twice as high as the warehousing industry's average) also the most injurious example.[2]

And then there are the behavioral and environmental downsides of the much-increased speeds of production, consumption, and disposal of materials and goods. This is true of countless readily discarded products with very short life cycles, but plastic waste has certainly been the most obvious global manifestation of this acceleration of material use. From one extreme to another: people in preindustrial societies lived for years, even decades, with many much-used and worn-out items of daily use, while in modern societies single-use plastics have become an unavoidable part of daily life, a mass-scale irrationality adding up globally to millions of tons of instantly discarded items. And as the Great Pacific Garbage Patch testifies, massive waste has now reached even the parts of the ocean far from the nearest shores.[3] And plastics are just one component of a global waste system whose rates of use and abandonment greatly surpass the speed of recycling.[4]

As for the limits to speed, they range from fundamental physical realities (the speed of light, the maximum speed of pistons, the sonic boom created by supersonic speed) to paying unacceptably high prices, be it in environmental or monetary terms (or both). Finally, I will address a fascinating phenomenon of stagnating and even declining economic productivity and its links to speed. This new techno-economic reality has received a great deal of attention, but its main cause remains disputed. Looking back at speed and productivity, Canadian economist Bernard Beaudreau concluded that: "Metaphorically speaking, speed was the goose that laid the golden eggs over the past two centuries, producing an unparalleled increase in material wealth . . . However, like any goose—even those that lay golden eggs—its period of egg-laying came to an end in the late 1960s/early 1970s."[5]

We will see if this single-factor explanation is supported by available evidence—or have there been other factors involved?

Accomplishments: from products to the death of distance

Speed has always been a notable component of the quest for survival, requisite for evading natural dangers, killing prey, and accomplishing lifesaving tasks. And then there is the desire for speed as a much-appreciated part of entertainment, and a contributor to notable, constructive—and destructive—achievements. During chariot races, a part of the Olympic Games from 680 BCE, horse teams pulled drivers in two-wheeled carts around an oval track with hairpin turns, as fast as the drivers could whip them.[6] Equine successors of these races have survived into the 21st century but in terms of spectator attendance they have been, as already noted, vastly surpassed by motorized versions—car racing on oval tracks, winding roadways, city streets, and across rough terrain.

Examples of remarkable speeds during antiquity range from rapid forced marches to spanning major rivers. Alexander's army, in pursuit of Darius III, was claimed to cover about 58 kilometers in one day, a remarkable speed for such a large formation.[7] Even more astounding is the speed with which in 55 BCE the Romans bridged the Rhine downstream from today's Koblenz, where the river was up to 9 meters deep: Caesar's engineers and 40,000 soldiers built a sturdy wooden bridge in just 10 days, a task that we would be hard-pressed to match with modern tools (and certainly unable to do when driving wooden piles into the riverbed by using nothing but heavy stones as hammers.[8]

A Roman charioteer whipping his horses around Rome's Circus Maximus: the empty space is still there, but the track and the three turning posts (*meta*) are long gone . . .

Half a millennium later, Byzantine architects and workers set an even more impressive record of speedy construction by building Constantinople's enormous Church of the Holy Wisdom (the original name of the massive and resplendent structure known as Hagia Sophia) in a mere five years, between 532 and 537.[9] But these were admirable exceptions of construction speed. As already noted, many monumental buildings required centuries to complete, and those exceptionally capable military leaders and ingenious architects had no contemporaries who could devise faster travel or rapid mass-scale manufacturing.

The best we could do to speed up the movement of heavy loads was to multiply the labor units and optimize the organization of such moves to harness the animate energies of laborers and animals. Domenico Fontana's raising of the Egyptian obelisk in 1586 is perhaps the best illustration of complex task coordination, while the ancient Egyptians figured out a helpful practice: notice the man in front of the statue-carrying sled pouring water on the sand in the nearly 4,000-year-old tomb painting below. Modern experiments have shown that the sliding friction of sand is greatly reduced by adding water—not too little, but also not too much. Adding some water improves sliding; adding too much impedes it![10]

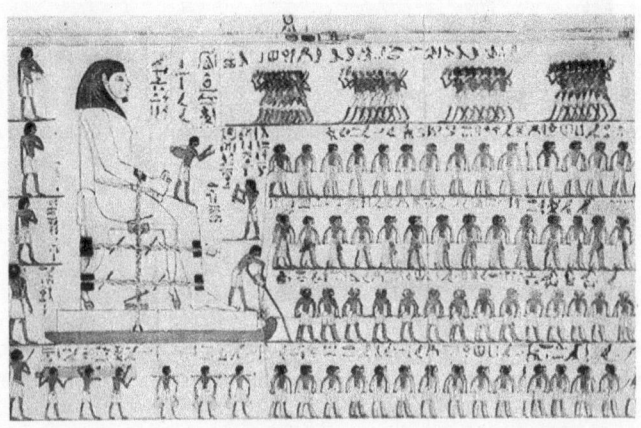

1880 BCE, a wall painting from the tomb of Djehutihotep.

How we have advanced

Stagnation and slow speed gains marked the long premodern and early modern periods. On land, harnessed oxen or water buffalo did not work faster in the last century of the early modern era (1700–1800) than they did a millennium earlier—but the Dutch windmills of the 17th century could reclaim land from the sea much faster than the earliest medieval designs of wind-driven machines that appeared during the 8th century CE in what is today's Iran. At the height of their power, the Romans were crossing only the Mediterranean (*mare nostrum*), hugging the shores of Atlantic Europe, and making forays into the Black and Red seas: they had no means to sail across open oceans to other continents.

This era of speed stagnation and slow incremental advances ended only with the large-scale commercial conversions of new energies (fossil fuels) during the first half of the 19th century.[11] After decades of slow adoption, steam engines were deployed in land and water transportation starting in the 1830s, and in the 1840s the telegraph brought a fundamental advance in the speed of communication. The second stage of this rapid modernization took place before the First World War with the invention of internal combustion engines and electricity generation, and with the production of crude oil followed by much-improved ways of its processing to yield portable fuels of very high energy density.

Those unprecedented developments resulted not only in higher machine and tool speeds (L/T) but also in processes that could be operated continuously (or at least for extended periods of time before the needed servicing)—hence greatly raising the overall output (N/T) speed, potentiating the impact of higher working speeds, and resulting in impressive 19th- and early 20th-century gains in industrial productivity.[12] The widespread adoption and use of high-speed machinery enabled faster and more affordable travel, while the availability of cheaper electricity and natural gas sped up all industrial, commercial, and household tasks.

The last, still unfolding, stage of increased speeds has been driven by new and improved prime movers (internal combustion engines, including giant diesels and gas turbines in flight; electric motors for high-speed trains) and, above all, by enormous advances in solid-state electronics. Their soaring performance and falling costs have led to the ubiquitous deployment of computers in all economic sectors, and eventually to the rise and rapid global adoption of truly global voice and data transfer, and instant, inexpensive mobile telephony.

Quantifying the past trajectories of speed means contrasting long periods of no or small speed increments and the steep to near vertical rises of the past two centuries, which have often been followed by sudden transitions to maximum or near-maximum levels. Time to prepare a hectare (100 × 100 meters) of land for seeding declined from 500 hours for hoeing to about 20 hours for working with weak oxen during European antiquity, then to about 10 hours for a horse team and to less than 2 hours for a multi-share plow pulled by a tractor.[13] This leaves little room for further speed gains in today's fastest rates of tillage. For decades, farmers in affluent countries have had an option to practice zero-tillage cultivation thanks to herbicide applications and direct planting into undisturbed soil.[14] But this option is not practical for most food crops in most environments, and hence its further extension (even if it met no environmental objections) will remain restricted.

The time required to produce a kilogram of wheat, the staple grain of Western civilization, shrank from about 25 minutes in early imperial Rome to 18 minutes a millennium later; by 1800 the most intensive (relatively high-yielding) European practices would need only 5 minutes; and by 1900 relatively low-yielding but highly mechanized American methods required no more than 1.5 minutes. Machines powered by internal combustion engines (tractors and combines) shrank that time further, to less than 30 seconds by 1950 and to a mere 2–3 seconds by 2020: with the help of metals (steel, aluminum, copper) and other materials (rubber, plastic, glass) to build engines and machines, and with the combustion of fossil fuels to power them, the speed of wheat production increased 15-fold during the 20th century and more than 160-fold since 1800. Further N/T speed gains would most likely come from higher yields rather than faster operating speeds.

Preparing wheat harvests for consumption by grinding progressed from manual querns producing 2–3 kilograms of rough flour per hour to 7 kg/h for rotary mills operated by people and 10–25 kg/h for the larger version turned by animals. Roman water mills could produce 100–200 kg/h, and by the end of the early modern era their best versions could turn out more than 500 kg/h. In 1786, England's first steam-powered mill, Albion Mills, could produce 3,400 kg/h; by 1881, the Pillsbury A-Mill in Minneapolis, at that time America's largest, could mill 15,000 kg/h; and today the largest US facility, the North Dakota Mill, can produce more than 350,000 kilograms of durum flour in a 10-hour shift.[15]

Steel is perhaps the best indicator of a civilization's long-term material progress, but it remained difficult to produce and was reserved for special uses until the late 1850s.[16] The global annual steel output rose from thousands of tons during the High Middle Ages to tens of thousands of tons before 1800 to about 200,000 tons in 1850, then to 28 million tons in 1900, 850 million tons by the year 2000, and to just over 1.8 billion tons in 2022—a gain of about five orders of magnitude (100,000-fold) in 250 years. A better indicator is to express this in per capita terms, with the average global supply rising from less than 200 grams in 1850 to nearly 18 kilograms in 1900 and to about 225 kilograms in the year 2020.

Because most of these productive advances took place during the 20th century, the simplest way to appreciate them is to calculate output multiples between 1900 and 2000.[17] For the primary energies these were 4.7 for coal (as its extraction was already considerable by the 1890s), but 199 for crude oil and 538 for natural gas, translating to a roughly 15-fold rise of the total fossil fuel supply. For the key materials the multiples were 30 for steel, more than 60 for phosphatic fertilizers, more than 100 for cement, and about 3,600 for aluminum. We cannot calculate the multiples for ammonia (its commercial synthesis began in 1913) or plastics (they took off only during the 1930s), but the respective multiples for 1950–2000 are about 30 and 120.

As for transportation, historical land speeds have remained constant for walking: a fast walk has been always about twice as fast as a normal slow pace (4 vs. 8 km/h). Distances of 25–35 kilometers can be covered comfortably in a day, but what mattered most in long journeys lasting weeks and months was their daily average, and that could be as low 3–4 km/h. The typical speeds of small and poorly fed pack animals and oxen harnessed to carts and wagons moving on muddy and bumpy roads (the Roman stone-paved *cursus publicus* was an exception, not a rule) were no better, but horses, particularly when deployed in well-organized messenger services, would average 12–17 km/h—that is, traveling often in excess of 100 km/day—and these top daily courier speeds did not see any major gains for more than two millennia.

Only steam-powered trains brought a radical acceleration of land travel, with even the earliest attempts of the 1830s averaging at least 30 km/h. Maxima around 100 km/h were reached within a generation, but until the First World War typical speeds on scheduled intercity runs averaged mostly between 50 and 80 km/h. This extended the daily maxima of land journeys by the best trains to well over 1,000 km/day by 1900—a distance that is now covered in little more than one hour by commercial jetliners. The streamlined diesel locomotives of the 1930s could reach a maximum of 200 km/h and could sustain speeds above 120 km/h. These performances were surpassed only by the introduction of Japan's Shinkansen, the first rapid train designed for a speed of 210 km/h, while the latest models of electric trains, now operating in the EU, Japan, and (most extensively) in China, have scheduled speeds of 300–350 km/h.

Road vehicles could never compete with trains over longer distances either in terms of speed (except for Germany's *Autobahnen*, freeway speeds are limited to 100–140 km/h) or comfort and safety (with trains offering carefree rides that allow for sleeping, reading, or walking). Of course, cars can considerably speed up intracity trips and, in the absence of public transport, are the best choice for shorter intercity travel, but because of the need to regulate speed on public roads there have been only minimal historic speed gains. During the 1920s, the Ford Model T could do 65–70 km/h, faster than the speed now achievable on average within most American or European cities.[18]

On water, the speeds of oared vessels during antiquity were limited to 5–7.5 km/h, while those of Roman sailing ships were mostly 7–11 km/h with favorable winds and just 3–4.5 km/h with contrary flows. European sailings during the early modern era made longer-term averages no higher than 15 km/h, and the best sailing designs of the mid-19th century could briefly average (hourly or during an exceptional single-day run) 25–40 km/h. By the 1840s, paddlewheel steamships averaged 22 km/h for the complete crossing, and by the 1880s speeds around 35 km/h were common.

During the first decade of the 20th century, large ocean liners had speeds above 45 km/h, and the top post–Second World War perform-ances, just before steamships were made obsolete by jet engines, topped 60 km/h. Commercial aviation took off only after the First World War, with the first open-cockpit planes carrying a few passengers at about 200 km/h. Before the Second World War, mono-coque all-metal designs and more powerful engines raised cruising propeller-plane speeds to 300 km/h; the best post–Second World War designs (including the Lockheed L-1049 Super Constellation) reached cruising speeds of around 550 km/h before they were replaced by jetliners whose cruising speed of close to 900 km/h has remained unchanged since the late 1950s.

Half a millennium of successive speed gains can be traced very accurately by plotting the times required to cross what has become one of the busiest transportation corridors in history, the North Atlantic between Europe and North America. For centuries the westward crossings were at the mercy of prevailing westerlies whose fluctuating speed could prolong a journey by weeks. In 1584, during the first attempt to set up an English colony in North America, cap-tains Philip Amadas and Arthur Barlowe took 69 days to reach the coast of today's North Carolina.[19] Nearly 150 years later, in 1726, when Benjamin Franklin was returning from his first visit to Eng-land, the journey took 83 days, or nearly 2,000 hours.[20]

By 1838, steam-powered paddlewheel boats cut that to less than 16 days, large liners halved that time by the 1880s, by 1900 the lar-gest steamships took less than 6 days, and the pre–Second World War liner record (*Queen Mary* in 1938) was 3 days and (not quite) 21 hours (93 hours in total). The Lockheed Constellation made it in a bit more

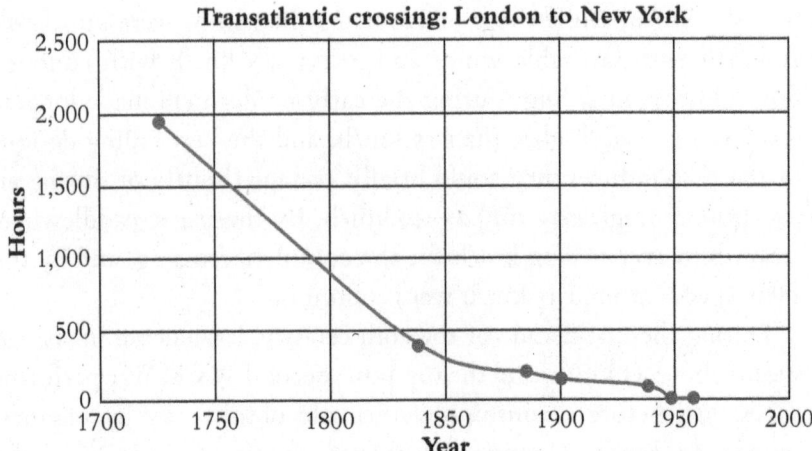

Transatlantic crossing: London to New York

Crossing the Atlantic: from sails to jet engines—from nearly three
months to just eight hours.

than 11 hours, and since 1958 westbound jetliners are scheduled to
arrive in just 8 hours (and with favorable prevailing winds in less than
7 hours in the opposite direction).

But, as impressive as this travel speed gain has been, the crossing is
also a perfect example of the inadequacy of simple quantitative speed
comparisons.

In Franklin's time, even the richest travelers could not buy real
comfort during the lengthy crossing, and experiencing a mid-Atlantic
storm on a small wooden sail ship was often a truly frightening event.
In contrast, crossing the Atlantic on a late-19th or early 20th-century
liner in first class could be (as vicariously appreciated by generations
of Europeans and Americans familiar with the tragedy of *Titanic*) a
luxurious experience, and even the poor emigrants traveling as third-
class passengers could access the deck and enjoy the life-changing
voyage. In contrast, most of today's rapid crossings by air take place
in the airborne equivalent of immured steerage class: bearable (unless
there are long delays or unscheduled landings caused by unruly passengers) but hardly memorable.

What all those new, post-1830 transportation speeds have accomplished can be shown in the most immediate graphic way by using

isochrones, the lines on a map that connect places that can be reached within the same specified length of time: as speeds have increased, accessibility has greatly improved and new transportation modes have resulted in a shrinking world. In the early 19th century, the fastest option was to travel by a post coach. Isochrone maps for 1819 show that coaches starting in Berlin (averaging 6–8 km/h) could reach Leipzig in 24 hours and Stettin on the Baltic Sea in 20 hours. By 1906 trains could reach Stettin in just 2 hours, and in 20 hours travelers could be in Amsterdam, Copenhagen, or Vienna. And by 2015 travelers could be in Leipzig in 2 hours, and in the 24 hours it would have taken to reach the city in 1819 they could be, by train, in Moscow.[21]

Isochrone maps, popular in the 19th century, are now available for many places and many years. In 1857, a traveler (whose fastest option was going by train and then by a coach) starting in New York could reach the border of North Carolina and the southernmost shore of Lake Erie in one day, and the eastern border of Texas and central Minnesota in one week.[22] North Carolina can now be reached from New York by car in eight hours, and by plane in less than two hours. In the 1850s, it took a month to travel from coast to coast; in 1930 it took a day of flying; now it takes six and a half hours.

The shrinkage of land travel has been most impressive in China. In 1949, the year of the establishment of the People's Republic, steam locomotive–powered trains, averaging 40 km/h, took 36 hours and 50 minutes to reach Shanghai from Beijing; now, the fastest sched-uled rapid train (No. G1, reaching up to 350 km/h) has cut the travel time to just 4 hours and 29 minutes![23]

Both freeways and rapid trains create distinctly tentacular iso-chrones, as cities within 100–300 kilometers can be reached by nonstop drives or rides in one hour—a shorter time than is required to get to many less distant locations that can be reached only by much slower peri-urban car drives and trains limited by posted speeds. These ten-tacular patterns become even more pronounced with air travel, as hub-to-hub flights (New York to London, Los Angeles to Tokyo, Dubai to Singapore) make many large cities rapidly and readily (with many daily services) accessible, even as getting to many smaller and much less distant destinations may require more time due to infrequent connections.[24]

The overall gain in the maximum speed of communication is easy to compare, from the speed of sound (343 m/s) to the speed of light (and electricity, nearly 300,000,000 m/s). That is very close to a million-fold gain, in speaking (or shouting) terms, as if instead of being able to reach somebody 30 meters away you could now communicate with somebody 30 million meters (30,000 kilometers) away: that would be like shouting eastward along the equator on Indonesia's Kalimantan and being heard, within a second, in Uganda—three-quarters of the equatorial circumference in the counterclockwise direction! This is not just a shrinkage of distance; it is a virtual elimination of distance as a factor in communication, and this is the new reality that was celebrated and commented on with awe during the mid-1990s as the Internet and mobile telephony ushered in the era of instant and inexpensive telecommunication.[25]

The speed of calculations has increased so rapidly during the past 150 years that it is far better to compare it just as orders of magnitude of floating-point operations: the sequence rises from one for mental calculations to hundreds for the first (mid-1940s) computers using vacuum tubes, thousands for the first transistorized machines a decade later, millions during the 1960s (with integrated circuits), billions two decades later, trillions by the year 2000, and a quintillion in 2022—altogether an increase of 18 orders of magnitude.[26] In relative terms, this is as if an organism able to crawl 4 millimeters per second was, just a few decades later, able to reach Proxima Centauri, the closest star to the Sun, 4.24 light-years (or 40.08 trillion kilometers or 40 quintillion millimeters) away. Obviously, no other speed gains come even close to those of number-crunching!

The worldwide premodern economic growth rate can only be estimated, but many careful reconstructions for countries and regions based on proxy records (harvests, prices, taxes) make it clear that pre-1600 growth rates were overwhelmingly either too small to be noticed or they amounted to small fractions of 1 percent. Post-1800 acceleration of economic growth began slowly as it was limited to a small number of regions, but it quickened after 1850, and, despite two world wars and an intervening world economic crisis in the 1930s, proceeded at the same heightened rate during the first half of the 20th century: the average global per capita GDP (with values in constant monies to eliminate the effects of inflation) rose about 1.4 times between 1800 and 1850, about 2.3 times between 1850 and 1900, and 2.4 times between 1900 and 1950.[27] That was followed by a quadrupling between 1950 and the year 2000, and during the first two decades of the 21st century the average global per capita GDP rose 1.4 times.

And there is another fundamental way to buttress the economic data and confirm the unprecedented speed of post-1850 development: by turning to the trends in global energy consumption. Comparisons are usually done in terms of gross energy production (of fossil fuels and primary electricity), but a more revealing way is to use the best estimates of final energy consumption (heat, light, motion) that account for changing energy conversion losses. Even during the closing decades of the 19th century, typical energy conversions were inefficient: less than 2 percent of electricity was converted to light by incandescent bulbs, no more than 5 percent of coal was turned into the kinetic energy of locomotives, and only about 20 percent of coal burned in small stoves actually heated rooms.[28]

In contrast, today's best natural gas–fired turbines generate electricity with efficiencies of more than 60 percent, large diesels convert 50 percent of their fuel into the kinetic energy of container ships or locomotives, and my house is heated by a 97 percent efficient natural gas–fueled furnace. As a result, the overall weighted efficiency of global energy use rose from only about 10 percent in 1800 to 15 percent in 1850, 20 percent in 1900, 30 percent by 1950, 45 percent in the year 2000, and 49 percent in 2020.[29] This means that while the world's total energy supply rose roughly nine-fold during the 20th century, steady improvements in conversion efficiencies resulted in a 20-fold rise of useful energy! In per capita terms, the availability of useful energy grew nearly five times during the 19th century, and it quintupled again during the 20th century.

That is almost identical to the growth of average per capita gross economic product, but for the second half of the 20th century the speed of average per capita GDP gains (three-fold) was faster (even when expressed in constant monies) than the speed of expanding per capita supply of useful energy (2.3 times). This could be seen either as a welcome shift—the declining dependence of economic growth on overall energy use—or as an artifact of the excessive valuation of services, now by far the largest sector of modern economies and one that requires much smaller energy inputs than agriculture, mining, and construction, without whose contributions we would not have any prosperous economies. This accounting disparity is particularly glaring when contrasting agriculture with the financial sector: in the early 2020s, the latter's contribution to global economic product has been more than four times that of producing food! But, obviously, losing the world's entire (and energy-intensive) food output would have consequences immeasurably greater than losing a quarter of today's financial services![30]

The need for speed—and its obverse

All the just-recounted accomplishments, all those admired widespread quests for speed, must have their obverse sides. The famous Latin oxymoron of making haste slowly—*festina lente*, translated from the less often quoted Greek original σπεῦδε βραδέως and chosen as a motto by Augustus, Rome's first emperor, and by Cosimo de' Medici, founder of the most famous of all powerful Italian Renaissance families—is nothing but a cautionary reminder based on a rich experience of problems, mishaps, failures, and outright catastrophes engendered by unwisely chosen speed: a warning about paying a price for being too fast.[31] The remit of this reality ranges from the already noted (in the second chapter) phenomenon of "running faster causes disaster" to (and here I deliberately choose an example very few people would think about) harvesting grain.

As for the first trade-off, losing control—especially in turns—can happen to every runner, be it an escaping or pursuing animal or an experienced athlete. Perhaps the most witnessed instances of this rashness are 1,500-meter runs, where the athletes go immediately for the inner lane and run most of the race in a clump where every sudden acceleration aimed at overtaking a rival can end in the fall of one or several competitors.

As for the latter, we have careful studies to show the economic effect of harvesting too fast: when combining winter wheat or winter barley, the speed should be kept at 4–6 km/h to limit grain losses to 0.3 percent for wheat and 0.6 percent for barley; when combining at a speed of 8 km/h those losses become unacceptably high, respectively tripling to 0.9 percent and doubling to 1.2 percent.[32]

In countless instances of human mental exertions, the *festina lente* phenomenon is known as the speed-accuracy trade-off (SAT). A pioneering study of "response amplitude and terminal accuracy on 2-choice reaction" was published in 1964, and the relationship shows an expected diminishing return in accuracy with additional time.

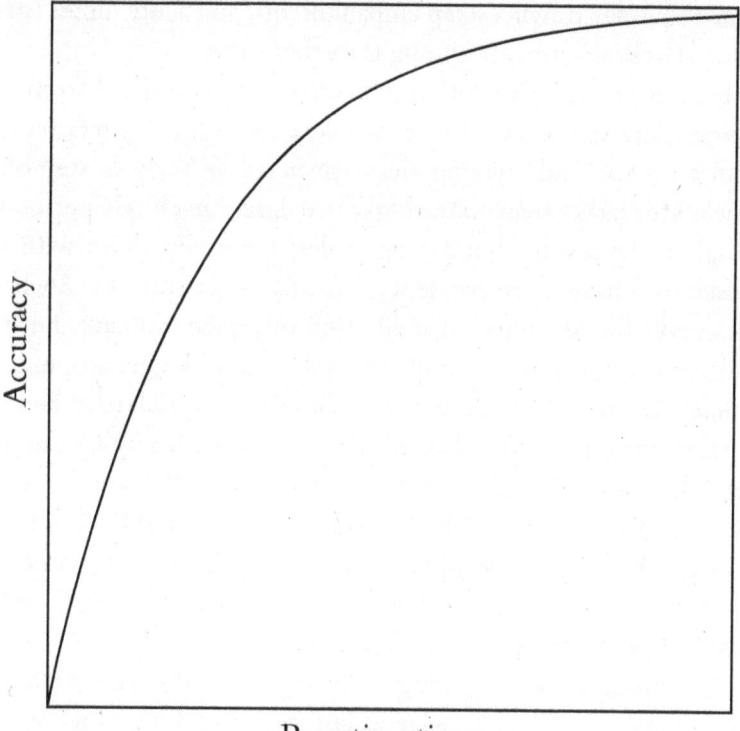

The idealized relationship between speed (reaction time) and accuracy.

Subsequent SAT investigations have included experiments with both people and animals, and recently they have been driven largely by the now-ubiquitous man-machine interactions between computers and people.[33]

Of course, by far the most tragic category of SAT is the outcome of excessive car speeds, when drivers lose the accurate options (negotiating a curve, braking within the required distance, avoiding a pedestrian or another vehicle). Given the global extent of fatalities caused by motor vehicle accident—1.35 million deaths a year, about twice as much as the annual global mortality due to breast cancer—it is worth taking a closer look at their relation to speed.[34] Obviously, speed is not the only cause of vehicle crashes: mechanical malfunction (now relatively rare), the state of the road, collisions with detached cargo or lost tires, distraction (which can be deadly even at low speeds, for example when a brief moment of inattention causes a car to plunge down a steep embankment), and acute illness (stroke, heart attack, seizure) are among the other causes.

But, in practice, it is often difficult to separate speed from other factors. An experienced driver can take a curve at a much faster speed than a novice, and a young alert driver is less likely to step on the accelerator rather than on the brake pedal than an elderly person with incipient dementia—but young males, especially those with male passengers, have more accidents than any other cohort.[35] Such realities make the unequivocal attribution of crashes difficult, but three realities are indisputable: higher speeds require longer stopping distances; the risk of a crash increases with higher speeds (that being the obvious reason for posted speed limits); and (basic physics cannot be denied), higher driving speeds lead to higher collision speeds resulting in more severe injuries, as higher kinetic energy must be transformed into much higher mass distortion and heat due to the inexorable proportionality of kinetic energy to the mass of the object and the square of its velocity (kinetic energy = $0.5mv^2$).

This means that, everything else being equal (the same vehicle, the same mass), a car going twice as fast will have four times as much kinetic energy to transform destructively at the point of contact. The effect is unmistakable even at relatively low speeds. An analysis of

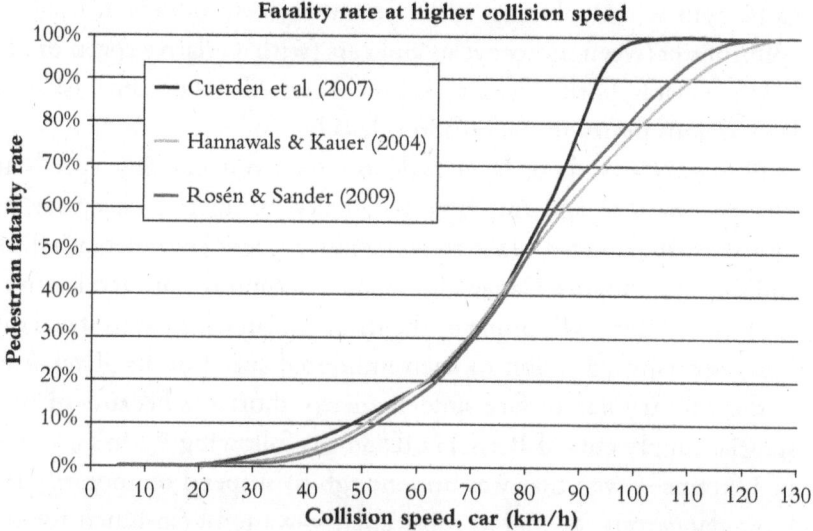

Fatality rate at higher collision speed

How speed kills: an exponential increase in the pedestrian fatality rate.

hundreds of crash tests done by the US National Highway Traffic Safety Administration showed that increasing the crash speed from 25 km/h to 56 km/h quintuples the risk of a severe injury from just 2 percent to 10 percent.[36] Pedestrians are, of course, the easiest victims of this energy-transformation disparity: their fatality rate is less than 20 percent when hit by a car going at 60 km/h, but more than 80 percent when colliding with a car traveling at 100 km/h.[37]

Seat belts (the three-point design introduced for the first time in 1959 in Sweden) have been great lifesavers: they have been very effective in nearly eliminating fatalities in speeds of less than 60–70 km/h. They provide less protection at speeds above 100 km/h, however, and even belted passengers in small sedans are at greater risk during collisions with the latest generation of fast and heavy vehicles (with some SUVs and pickup trucks weighing more than 2 tons).[38]

Motorcyclists are more vulnerable than car drivers and passengers. A study of more than two decades of all German motorcycle injury crashes showed a strong and significant relationship between relative speed and injury severity: at 70 km/h, the risk for at least serious injuries was as high as 51 percent when hitting crash barriers, and

64 percent when colliding with narrow objects; moreover, head-on collisions between motorcycles and cars (with a relative speed of 120 km/h—that is, both objects at 60 km/h) carried a 55 percent risk of at least serious injury to the motorcycle rider.[39]

Not only has there been little progress on reducing speeds in Europe, the US, and Australia, but many jurisdictions keep raising limits: in 2021, seven US states introduced new laws to that effect, and in 2022 Ontario, Canada, raised its maximum from 100 km/h to 110 km/h.[40] But no example of the irrational attachment to speed can beat Germany's decision to keep unlimited speed on its *Autobahnen* as the country had to face sudden energy shortages because of substantial supply cuts of Russian oil and gas following Putin's invasion of Ukraine—even that was not enough to suspend temporarily the most dangerous, as well as most energy-wasteful (so much for the endless professions of German greenery!), form of road travel.[41]

There is little need to make more comments on (the automobile's) speed's effect on modern cities. Even a cursory look through books on the history of urban planning, or a quick dip into the Internet's resources, finds repeated complaints of how cars ruined or drove to the brink cities of the 20th century, how all that space (not only roads, but parking lots and garages) reduced and fragmented areas where people could live and move around, even as those cars made their daily lives more noisy and more polluted.[42] Of course, an unintended consequence of these massively disruptive changes is that they sped up intraurban flows for a while (up to the permitted limit), but as the numbers of cars grew faster than the easily passable (without traffic stops) roads, average urban speeds began to decline, part of an escalation that has seen the expansion of roads and traffic—leading to congestion, increased energy use and air pollution, more traffic accidents, and higher infrastructure maintenance costs.

The vicious circle of congestion is that, despite building more roads to make room for more cars chasing more cars, delays are increasing. All kinds of solutions have been tried: synchronizing traffic signals, imposing tolls, restricting vehicle access to parts of a city or to certain registration numbers, promoting shared rides, introducing congestion pricing. But the trend has not been dislodged: in the

US, the average hours of delay per motor commuter have been rising linearly for decades—in LA from 60 hours in 1982 to 100 hours in 2019, and during that time they had more than doubled in New York. This is an indubitable confirmation of traffic going backwards speed-wise—and incurring economic and quality-of-life losses that add up annually to hundreds of billions of dollars.[43]

And as a frequent flyer I must note at least two air travel–related problems with speed. The first one is that the most modern and most efficient jetliners are designed for the most profitable long-distance routes between large cities, be it from New York to London or from Atlanta to Chicago. Europe's, Japan's, and now also China's networks of rapid trains have made even many smaller cities rapidly accessible, but in countries without dense and rapid train connections (North and Latin America, as well as Africa and most of Asia) the choice is between tedious driving or flying in what are usually older, smaller, slower, less comfortable airplanes that operate on infrequently (and recently also much reduced) schedules.[44]

Moreover, on shorter intracontinental or domestic flights, the advantage of jet speeds has been greatly reduced, and in some cases even erased, by the necessity to travel to and from sometimes distant airports, undergo security checks, and deal with increasingly frequent flight delays and cancellations. In many cases, an hour may be required to reach the airport, airlines now advise travelers to arrive two or even three hours before departure, and delays, waiting for luggage, and a car or bus ride to the destination may easily add at least another hour—turning a domestic two-hour flight into a seven-hour experience. That would reduce the de facto traveling speed to less than 250 km/h, easily equivalent to high-speed train travel—but, as just noted, in most countries that option is not available.[45]

And then there is jet lag, a universal speed-related downside. This temporary sleep problem, sometimes brief but for some people uncomfortably long-lasting, arises from the clash between a rapid displacement across time zones and every organism's circadian rhythm—an internal clock set to a 24–25-hour cycle that is synchronized to the original time zone.[46] The effect can take place by crossing just two time zones, though usually it is felt after moving across at least four time zones, and for myself and many other travelers it becomes debilitating when midnight becomes noon—and a bit more. In my life I have experienced this dislocation at least 50 times when flying from Winnipeg in the geographic center of Canada to Japan (14 hours ahead) or China (13 hours ahead).

The usual consequences are feeling tired during the day, having difficulties staying awake and mentally alert, problems with digestion, and finding it impossible to fall asleep when the local time for sleeping arrives. There is a common belief that matters are a bit easier when flying eastward rather than westward, some studies found increased susceptibility with age, and there is no shortage of advice on how to minimize jet lag's effects (from drinking plenty of water to taking drugs). In any case, it is not the distance that matters; it is the speed at which it is bridged: at the beginning of the 20th century transatlantic liners took six days to cross from the US to the UK, losing an hour every 24 hours, which is a very manageable time change compared to losing an hour every 70 minutes when flying from New York to London.

Suffering jet lag when flying from Vancouver (with Olympic-grade downhill skiing in Whistler just a few hours' drive north of the city) for a week of skiing in the Alps (about 8,500 kilometers away) is a matter of (dubious) privileged choice. Trying to cope with increased speed in a workplace is another matter entirely, and in few workplaces has this peril been more common than in meatpacking. US work injury statistics show that the average annual rate of nonfatal injuries in animal slaughtering and meat production from carcasses (7.4 percent of the workforce injured in 2020) is nearly three times as high as in logging (2.6 percent), a sector well known for its work-related hazards due to the heavy machines and tools and the threat of large falling tree trunks.[47] Moreover, the specific tasks dominating the meatpacking industry (sawing, cutting, slicing, lifting heavy pieces of meat) cause the highest rate of work-related musculoskeletal disorders—seven times the average incidence rate in manufacturing.[48]

Dangerously fast slaughter-line speeds are the leading cause of worker injuries, due to the pressure to dismember more animals in less time. Even without knowing the details of the successive steps—from stunning by submerging animals in CO_2, bleeding, skinning, and eviscerating them, to cutting up and trimming carcasses, and producing the desired retail cuts—it is obvious that the entire process requires all kinds of specific motions that can easily lead to repetitive stress injuries of the hands, arms, back, and legs.[49] Think of spending the entire shift just piercing pig jugulars, or slicing skin off pig heads, or scraping hair from the skin. And given that so many tasks are done by workers standing along moving belts and wielding heavy knives, cuts are an ever-present risk, in extreme cases requiring amputation.

Speed exacerbates these risks: "The line is so fast there is no time to sharpen the knife. The knife gets dull and you have to cut harder. That's when it really starts to hurt, and that's when you cut yourself."[50] Just reading this is painful—and yet in 2019 a new US rule removed a long-standing hourly slaughter limit of no more than 1,106 pigs, and set no upper value as long as fecal contamination is prevented and bacterial counts are minimized. This means that some plants now require the slaughtering and processing of more than 1,200 and even 1,300 pigs per hour. Not surprisingly, these higher speeds of processing have resulted in higher rates of injuries and fatalities: in 2020, the Occupational Safety and Health Administration conducted 27 inspections at meat plants due to a fatality or catastrophe, compared to just one in 2017, three in 2018, and five in 2019.[51]

At the same time, it must be acknowledged that modern societies have benefitted enormously from increased speeds, with the Covid-19 pandemic providing numerous illustrations of such benefits. Vaccines were developed and administered in record time: the WHO proclaimed the new pandemic on March 11, 2020; the first mRNA (Pfizer) vaccine became available in December 2020; and vaccines were rapidly produced in unprecedented quantities (they also required record supplies of special borosilicate glass vials), and distributed worldwide by cargo jetliners. By April 2023, more than 13 billion doses had been administered.[52]

Speed limits: physical, economic, environmental

The speed of cellular and organismic growth is limited in many ways. On the most fundamental level, an extensive examination of about 4,000 bacterial proteins and 36 growth rates indicated that ribosome biogenesis is a primary determinant of growth rate in *Escherichia coli*, a common bacterium often used in laboratory investigations.[53] Ribosomes are structures inside a cell composed of ribonucleic acid (a large molecule indispensable for protein synthesis) and protein, and are the sites of protein synthesis. The growth of any organism, be it tree seedlings or humans, is inherently limited by its evolutionary heritage (piglets can grow faster than babies) and by the supply of nutrients: shortages cause stunting; and surfeits (above all, proteins in dairy products and meat) lead to populations of unprecedented, but still limited, heights—compare Dutch young adults with the citizens of East Timor.[54]

Inevitably, limits to organismic growth put limits on maximum physical exertions and competitive performances. Perhaps the most revealing examination of the speed limits of human performance has been provided by historical analyses of 10 outdoor sports events that are among the oldest and most popular in sports history.[55] The oldest contests include the Oxford vs. Cambridge boat race (since 1829), swimming across the English Channel (since 1875), and the hour cycling record (since 1893); other notable events are the Swedish cross-country ski race Vasaloppet (since 1922), the speed ski record (1930), and the Hawaiian Ironman Triathlon (since 1978). Exponential decay describes all of them: speed gains were fastest during the earliest phase (with the gains particularly pronounced for the oldest events), and are now approaching their asymptotic limits, which may be reached between 2016 and 2081, with the mean time in 2049.

As for other speed competitions testing human physical attributes, there is no stranger category than the many kinds of disgusting speed-eating contests: they prove only that the limits to stuffing oneself with assorted edibles (from hot dogs to the hottest-tasting peppers) are higher than one could reasonably imagine.[56] Similarly, the inevitable

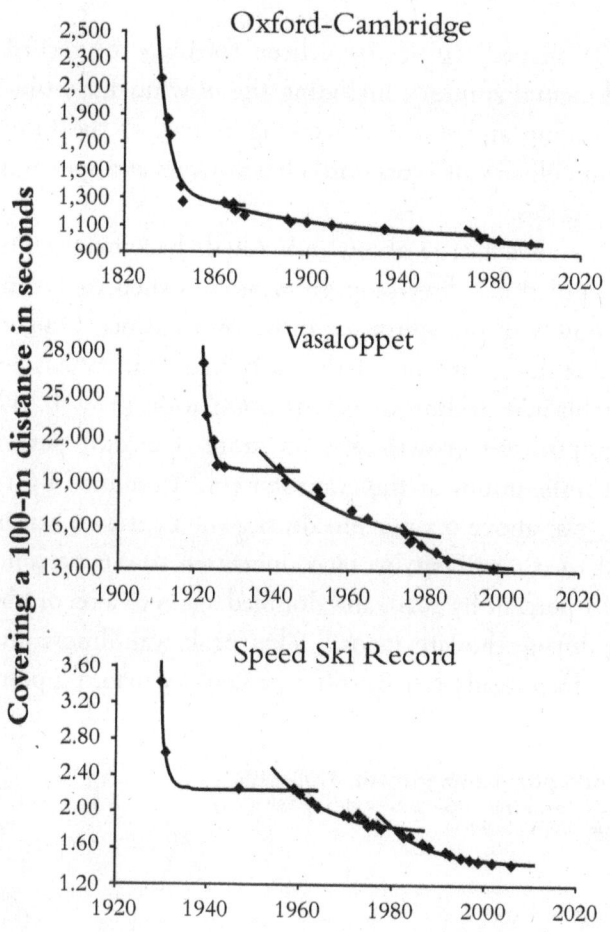

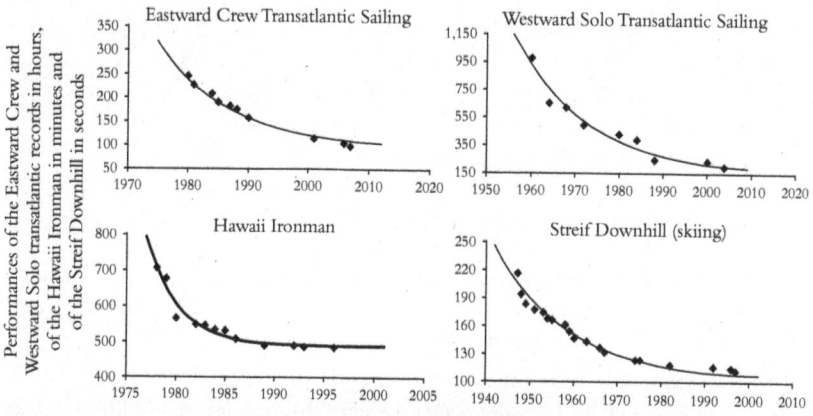

For human performance, faster is getting harder, be it running, rowing, or skiing.

growth (S-shaped) curves have been forming for record times of assorted mental contests, including the now quaintly old-fashioned manual typing and shorthand-taking as well as the more fashionable rapid delivery of crossword clue answers and speedy solving of sudoku puzzles.

What is always true about individuals becomes eventually true about populations: they stop growing and then they can stagnate, decline slowly or precipitously, or become extinct. One of the most remarkable developments of the early 21st century has been a now well-established decline of global population growth. The annual global population growth was no more than 0.05 percent during the first millennium of the common era; it doubled by 1500 to 0.1 percent, was above 0.2 percent during the next two centuries, then doubled to 0.4 percent by 1800, increased to 0.6 percent by 1900, reached 1 percent by 1940, and doubled again to a record level of 2.1 percent during the late 1960s.[57] That peak was almost immediately followed by a steady retreat to 1.7 percent by 1990, 1.4 percent in the

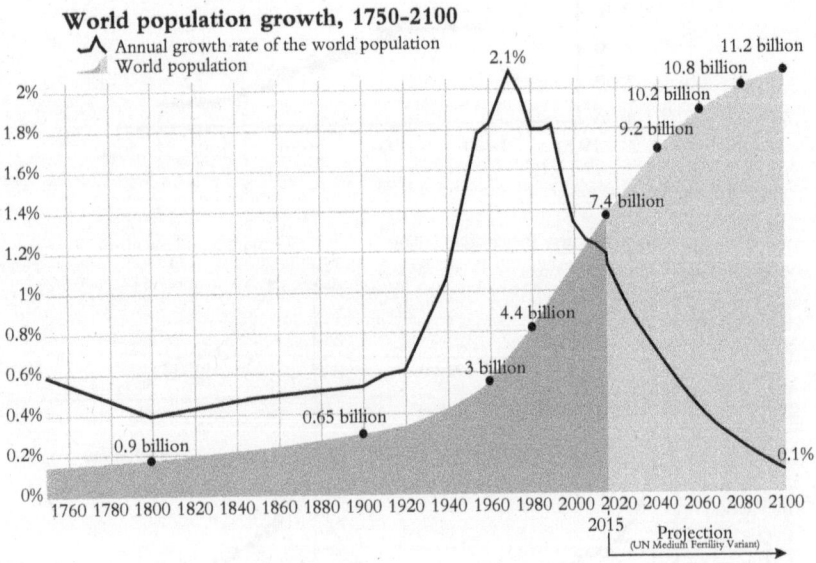

World population growth, 1750-2100

Reverse speed: during the 1960s, as the annual growth rate of the global population soared, many people saw no end to this rise; less than a lifetime later, many people worry about population retreat.

year 2000, and just above 0.8 percent in 2022, when it was no higher than in 1930. The speed of global population change, a key variable that affects many social and economic trends, has shown a clear sign of steep temporal rise followed by a similarly steep decline.

As a result, an S-curve of global population has been forming, and the latest data on total fertility rate (the average number of live births per a female lifetime) show that more than half of the world's countries (including China, India, Indonesia, US, Brazil, Russia, and all of Europe) are now below the replacement level.[58] In the absence of immigration, these fertility declines soon translate into national population declines, and shrinking nations cannot be expected to be as dynamic as those whose populations are still in the ascending phase of the growth curve. In 2023 the countries with declining populations included Japan, Italy, Ukraine, Poland, and Romania, as well as 14 other smaller European nations.[59] Depopulated villages are already common in Spain's Guadalajara, in parts of western Romania, and throughout northern Honshu in Japan. All too obviously, the tempo of life in the heavily depopulated villages of Japan (by 2033, 20 percent of the country's houses will be empty) is very different from that of Nigeria's expanding cities.

In depopulating regions, the perception of ever-faster human affairs could refer only to the speed (N/T) of population shrinkage. And even in areas whose populations are still growing, people are not walking faster than they were two generations ago—peering into mobile phones while walking cuts the speed and, overall, Americans have been walking less, a trend only strengthened by the pandemic.[60] Cars, commuter trains, and airplanes do not travel any faster than they did 40 years ago. In fact, congestion and delays have increased the likelihood that a particular ride or flight will take longer or not happen at all: 2023 was a year notable for a deterioration of comprehensive (door-to-door) travel speeds involving long-distance flying! Nor are most quotidian tasks—from cooking on an electric range to vacuuming a room—done faster (i.e. more dishes finished or more rooms cleaned per hour) than they were 40 years ago.[61]

Much like organisms and populations, machines have their own growth limits, and after more than 150 years of inventing, adopting, and perfecting an ever-wider range of powerful designs—no matter if they are driven by fossil fuels, electricity, or natural energy flows (wind and solar radiation)—we have now reached, or closely approached, many performance limits. This obviously invalidates the frequent claim that technical capabilities have been increasing exponentially, with higher speeds and shorter transition intervals.[62] That may have been true about computing capabilities and a few other innovations (DNA sequencing, additive manufacturing, neuro-technology, machine learning), but not about modern technical innovations in general. There are still opportunities for raising some typical speeds in nearly all economic sectors, but by now they have been distinctly outnumbered by performances that cannot be substantially improved. I will start this review of machine speed limits with long-distance travel, the most common ubiquitous speed experience.

Three classes of machines have revolutionized personal transport and become the principal contributors to the shrinkage of distance: motor vehicles on good-quality roads, trains, and airplanes. There is no need to belabor the speed limits of the first class of these energy converters: they (with the notable exception of many stretches of the German *Autobahnen*) are posted everywhere, ranging from just 30 km/h on city streets with schools to the typical 60–80 km/h for much intracity traffic, and to freeway maxima of 112.7–128.7 km/h (70–80 mph) in some US states.[63] These speeds could have been reached by mass-produced vehicles a lifetime ago, and the posted maxima are safety-related limits whose longevity makes it clear that no radical (and technically possible) gains await.

Between the 1930s and the 1970s, trains experienced decades of speed stagnation caused largely by the demise of steam power and by strong competition from road vehicles (boosted by very low pre-1973 gasoline and diesel fuel prices) and by more affordable flying. Japan, with its introduction in 1964 of the world's first rapid train, was a singular exception, basically doubling the maximum operating speed from around 100 km/h for steam- or diesel-powered trains to around 200 km/h for the Shinkansen. But a large-scale renaissance began only with the French TGV (since 1981) and with the adoption of similar designs in other EU countries, and it received a decisive boost with the Chinese decision to build the world's longest network of rapid trains (the first link constructed in 2003). These developments raised common operating speeds to 200–250 km/h and maxima to 300–350 km/h, notable gains that are particularly impressive when compared to limited road speeds.

But that speed boost has now also run its course. Over short and medium distances, rapid trains will remain superior to both road travel and airplanes, and because of the time saved on trips to and from airports and on airport security, they may be time-savers even on distances of 600–700 kilometers—but because of many limits on rail travel, their maximum operating speeds cannot be increased even by a third, that is to about 450 km/h. That does not mean that a rapid train cannot travel at such a speed, however. In 2007 SNCF, the French railway operator, spent 30 million euros on Operation V150— named after its target of reaching the record speed of 150 m/s, or 540 km/h.[64] The test was run by a purpose-built short trainset (three coaches between two locomotives acting in pull-push mode, with no passengers). The goal was surpassed: on April 3, 2007, at 13:13:40, the set reached 574.8 km/h on the TGV-Est line between Prény and Champagne-Ardenne, far surpassing the 1990 record of 515.3 km/h.

At the same time, the maximum speeds achieved by rapid trains during their daily runs have not gone above 300–350 km/h. What are the challenges, and limits, of going faster? Steel wheels must be in contact with steel rails (only magnetic levitation could dispense with that), and hence it would help to minimize friction forces between the two surfaces. But the contact area between the wheel and rail is already only about 250 square millimeters, and an even smaller area would make it more difficult to stop a fast-moving train in short distances.[65] Large trains consisting of many coupled carriages and requiring pantographs (the articulated arm that holds the strips contacting the overhead wires to draw electricity) cannot be made as aerodynamically sleek as an airliner's body. Moreover, air drag is directly proportional to the square of speed, making the challenge more taxing. Overhead wires should be as light as possible in order to be kept as taut as possible—but, at the same time, they should be highly wear-resistant.

And then there are the interactions of a rapidly moving train with its surroundings, with infrastructure in tunnels, with crosswinds in open areas (especially on viaducts), and the train's effects on populated surroundings. Noise and vibration are not easy to control: noise energy increases in proportion to the 6th power of speed, and because 86 percent of the 513-kilometer-long track between Tokyo and Osaka runs through residential and commercial areas where noise must be limited to 70–75 decibels, Shinkansen trains cannot travel at their maximum speed.[66] Reduced control margins at higher speeds, the need for even better signaling, wear-and-tear on the rolling stock that is expected to last for millions of kilometers, higher maintenance costs and increased energy consumption are additional reasons for keeping normal high-speed train runs below 400 km/h.

Similarly, more than six decades of unchanged cruising speeds of commercial jetliners make it clear (as already explained in the fourth chapter) that the near-sonic optimum at around 900 km/h is a limit that is not easy to surpass.[67] Although Concorde's supersonic flights (with limited frequency and on limited routes) lasted nearly three decades (between 1976 and 2003, with scheduled flights mostly from London or Paris to New York, and also to Washington, Miami, Bahrain, Singapore, and Rio de Janeiro), those 14 planes (there were also six prototypes) could never make a profit, and the sonic booms they created restricted their flight to routes above the ocean.[68]

And much like mobile machines, stationary designs have also reached, or have come very close to, their performance limits. I will start with the now-popular renewable energy conversions: wind- and solar-generated electricity. The speed of these energy conversions is limited either by fundamental physical laws or by profit- and safety-related considerations. Large wind turbines are an excellent illustration of both limits. Obviously, these enormous machines (with more than 200-meter-tall steel towers and more than 100-meter-long plastic/balsa blades) require a minimum wind speed to start rotating: they usually cut in when winds reach between 11 and 14 km/h.

Faster winds are desirable because kinetic energy increases with the square of the wind speed—a doubled wind speed will produce

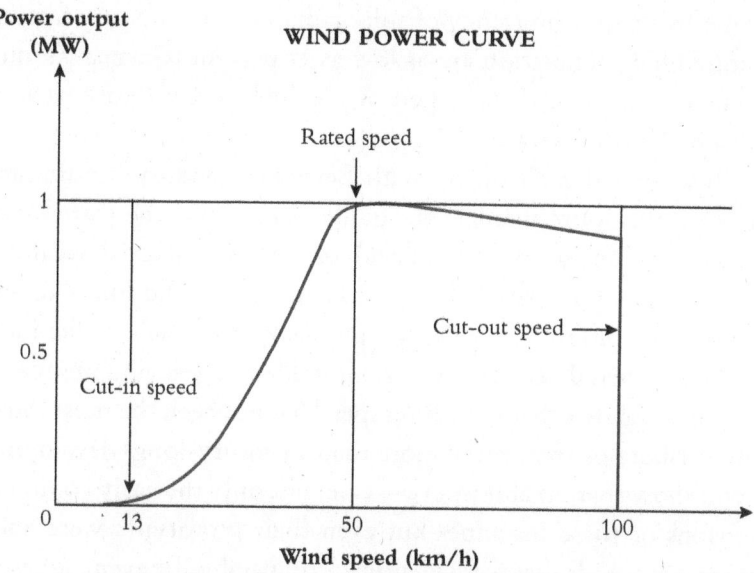

Cut in and cut out: the wind turbine speed window.

four times as much energy—but there are obvious structural limits to wind speed increase. Heavy storm (>100 km/h) and hurricane wind speeds (commonly more than 200 km/h) can rip the rotating blades off their nacelles, and they, in turn, can damage the tall steel towers. To guard against high wind-speed damage, all turbines have an automatic cut-out and get shut down once the wind reaches about 25 m/s (90 km/h), the speed at which some trees can be uprooted.[69]

And, as already noted in the third chapter, no matter how fast the winds are and how rapidly the blades turn, there is a limit to the efficiency with which any wind turbine can convert the kinetic energy of wind to electricity. Theoretically that is just shy of 60 percent, but in practice is no more than about 45 percent. Furthermore, the capacity factor (the fraction of time during which a turbine generates electricity) can be in excess of 50 percent in the windiest offshore locations, but in less windy onshore locations it is often no more than 25 percent, leaving the machines idle 75 percent of the time.[70] Meanwhile the conversion of solar radiation to electricity in photovoltaic cells is limited by the maximum theoretical efficiency of solar cells (the Shockley–Queisser limit of about 34 percent) and, as with wind, by the intermittency of solar radiation: the capacity factors of photovoltaic generation are as low as 11 percent (Germany's annual average) and more than 25 percent (in such sunny locations as Arizona or North Africa).[71]

How does this all square with Bernard Beaudreau's claim about speed as the goose that laid the golden eggs "over the past two centuries, producing an unparalleled increase in material wealth ... but that this potential is limited. That is, once the limits to speed are reached, no further gains are physically possible"?[72] The history of fuel-powered converters—those golden egg–laying engines and turbines—shows that increasing speed has not been the most important attribute of their (now more than a century-long) development. I will show that, in almost every case, not only the early commercial versions of these machines but even their prototypes were able to operate at speeds that were identical, comparable, or even higher than the speeds that now prevail in their everyday deployment. If absolute speed is the only indicator, then its increase has been, overwhelmingly, a matter of inventive saltation, not of gradual speeding-up.

These new energy converters delivered speeds superior to their animate or simple (water- and wind-driven) predecessors right from their earliest deployment—but their subsequent diffusion and eventually universal adoption did not take place solely because of any additional increases in typical operating speeds, but because higher speeds, attainable even by the first commercialized designs, have been greatly potentiated by subsequent advances producing a remarkable combination of welcome gains.

The list should start with increased power and torque. Power is simply energy per unit of time; torque is the rotational analog of linear force: instead of pushing and pulling, think of a rotation (twist) applied to a shaft.[73] But even as these machines have become more powerful, they require less (in some cases much less) material per unit of power (gram per watt) than their initial designs, reducing the need for expensive raw materials.

Moreover, these designs (without exception!) have become more efficient, and their reduced consumption of fuel and electricity has played a major role in reducing environmental pollution (above all, improving air quality). They have also become more reliable, with some engines able to run for thousands of hours before a major check-up, and years can now go by without a single catastrophic accident. And, inevitably, these advances have combined to lower the operating costs and deliver superior profitability. I will demonstrate this combination—of increased power, torque, efficiency, reliability, and profitability—by comparing the early models of these new, fast converters with the latest, superior designs.

The acceleration process started with steam engines. As is well known, they were invented and commercialized during the 18th century, but it is much less appreciated that they began to make a substantial difference, in both transportation and industrial production, only after the 1830s, when they provided the foundation of the new industrialized world as their reciprocating motions were transformed into rotary speeds as high as 300–500 rpm.[74] As we have seen, these speeds were fast enough to revolutionize land and water transportation. By 1900 we had the highest train speeds approaching 100 km/h, and large steamships could surpass speeds of 40 km/h on intercontinental voyages—and steam engines also provided unprecedented power for industrial processes, resulting in productivities that surpassed previously used manual processes often by orders of magnitude and resulted in a greater variety of more affordable household and industrial products, as well as in more diverse and cheaper diets.

But steam engines have inherently low efficiency (steam locomotives often converted as little as 5 percent of the coal they burned to motion, while the typical efficiencies of large stationary steam engines were no more than 15 percent), resulted in excessive environmental pollution, and occupied a great deal of space and were too heavy (their large interior volumes made them unfit for small-scale tasks and impossible for any airborne applications), and so they were rapidly replaced by better alternatives. The following comparison of steam engines and steam turbines designed for identical tasks shows why the former lost their place to the latter so rapidly.

Between 1902 and 1906, London County Council built a new tramway power station at Greenwich to supply the city's streetcars with electricity. Their choice of steam engine was what a historian of steam power called a "megatherium of the engine world."[75] The machine was 14.6 meters wide and 14.5 meters tall—while a steam turbine of identical capacity was just 3.35 meters wide and 4.45 meters tall.[76] Obviously, these giant machines required correspondingly large buildings to house them, and to convert reciprocating motion to rotations driving dynamos. This voluminous, massive, and complicated arrangement could do the job; the main advantage of the steam turbines that replaced these engines in 1910 was not their higher speed, but their smaller mass and volume and their higher efficiency and lower operating costs.

When followed from its beginnings to its most recent accomplishments, the sequence of inventing, perfecting, and adopting new machines is yet another excellent illustration of the universal weakness of a single-variable approach to explaining complex developments: speed was far from being the only or even the dominant consideration. Diesel engines, machines that I have called (together with gas turbines) the prime movers of globalization, are perhaps the best example of this surprisingly "speed-stable" reality. On February 17, 1897, Moritz Schröter—professor at the Technische Hochschule in Munich—ran the first official test of Rudolf Diesel's third

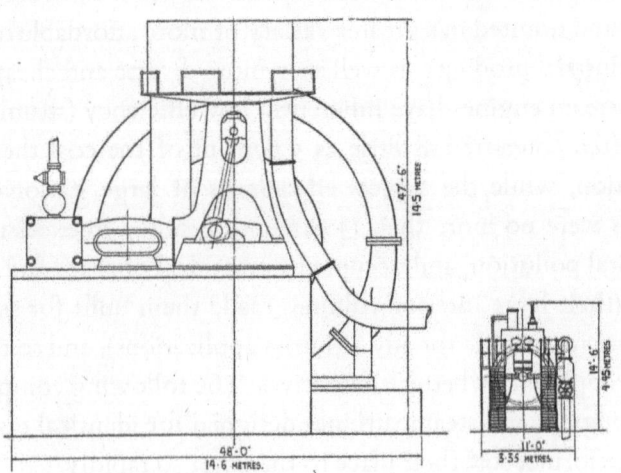

London's steam engine megatherium and an equally powerful steam turbine: the smaller mass, smaller volume, and higher efficiency—not the faster rotary speed—made the difference.

prototype engine and certified the following specifications: power 13.5 kilowatts, net efficiency 26.2 percent, and speed of 154 rpm.[77]

Now the world's largest diesel engine, the Wärtsilä-Sulzer RTA96-C (introduced in 2006), has a power of 80.1 megawatts, an efficiency just above 50 percent, and a speed of 102 rpm.[78] More than a century of diesel engine developments has culminated in a machine that is nearly 6,000 times more powerful and whose efficiency is nearly twice as high—but whose speed is a bit slower as it serves ship propulsion with propellers optimized for low speeds. Other large marine diesel engines have speeds mostly between 90 and 130 rpm. No higher speeds are needed to power their propellers: those should move as slowly as practicable; what those enormous vessels need is the unprecedented power and torque required to propel them and to maneuver them as they carry up to about 24,000 steel containers, more than 100,000 tons of bulk cargo, or the thousands of passengers and crew that make up the floating cities of large cruise ships.

The speeds of rotating shafts and propellers have not increased, but RTA96-C has a power and torque that are, respectively, nearly 6,000 and more than 20,000 times those of Diesel's 1897 machine! And what about the mass/power ratio? Diesel's 1897 prototype had a mass/power ratio of 333 g/W; the RTA96-C's ratio is 28.7 g/W: it needs about 90 percent less material per unit of power! Similarly, there has been no progression toward higher speeds as far as either steam turbogenerators or electric motors are concerned. Steam turbogenerators are the machines that used to produce more than 85 percent of the world's electricity during the early 1950s (the rest came from hydro turbines), and that even in 2020 were responsible for more than half of all electricity generation (with the rest coming from hydro, gas, and wind turbines, and PV panels).[79] I already noted (in the fourth chapter) how Charles Parsons's first small steam turbine prototype had an astonishing speed of 18,000 rpm and how it was rapidly reduced to 4,800 rpm for the first commercial application in 1891.

And while Nikola Tesla's first small polyphase motor, conceived and built in 1883 and 1884, ran at 11,800 rpm (and had a power of less than 200 watts), the speed of modern electric motors is selected by engineers according to the specific needs of the desired applications—with, once again, power and torque being the leading concerns. As a result, the world's largest electric motor, developed for the US Navy to propel ships, weighs 75 tons and has a power of 36.5 megawatts, but it works, with high torque, at only 120 rpm.[80]

The final speed comparison I will make is for gas turbines—now the world's most efficient electricity generators, whose unequalled performance is further enhanced when they are run in a combined cycle, using hot gas emitted by the gas turbine to heat water for a steam turbine, an arrangement that can boost overall efficiency by nearly 50 percent.

The first gas turbine, installed in 1939 in Neuchâtel, Switzerland, by Brown Boveri, had a net power of just 4 megawatts, achieved an efficiency of 17.4 percent, and turned at 3,000 rpm.[81] More than 80 years later, the Siemens SGT 5-9000HL offers a power of 593 megawatts and an efficiency of more than 43 percent in a simple cycle and 63 percent in a combined cycle, and it rotates at 3,600 rpm—just 20 percent faster than the pioneering machine.[82]

Few people see these turbines in operation, while everybody has seen gas turbines (jet engines) powering long-distance flights. Today's most powerful jet engines have a thrust (force generated by accelerating a mass of hot gas) 23 times higher than the engines that powered the British Comet, the world's first (and failed) commercial jetliner that entered service in 1952, but they are much lighter: their thrust-to-weight ratio is 2.4 times higher.[83] And gas turbines in flight also provide perhaps the most impressive illustration of the longevity and high reliability of the latest designs. Depending on the number of takeoffs (most done by airplanes on short intercity routes), well-maintained engines can be in service for up to 30,000 hours before they are overhauled, and they work so reliably that pilots will face in-flight engine shutdown less than once for every 100,000 hours in flight—the equivalent of an engine failure once every 30 years.[84] Obviously, failures due to engines ingesting foreign matter (most often bird strikes, but also volcanic ash) are more frequent.

Jet-powered flying also illustrates the large economic payoffs of

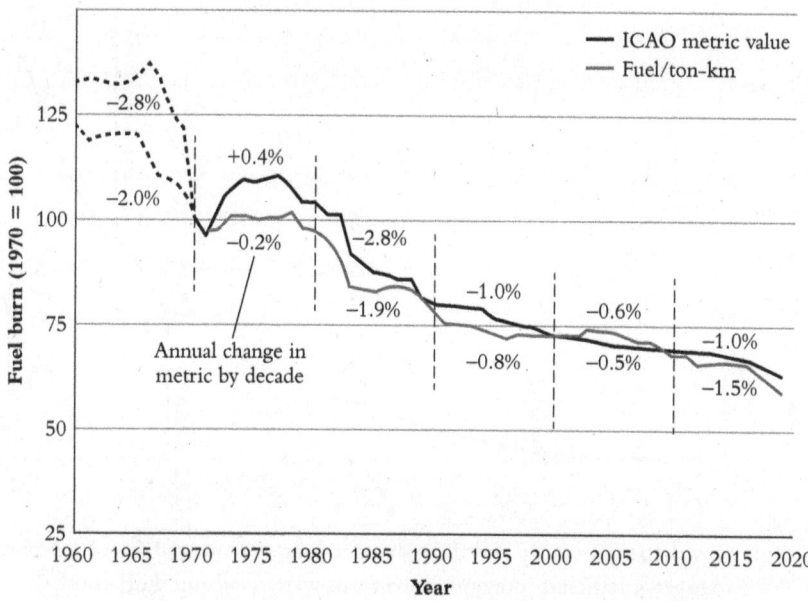

Speed is not everything: decades of near-constant cruising speed have been accompanied by steady efficiency gains, resulting in declining specific fuel requirements.

these combined, beyond-speed performance gains: the Boeing 787, the latest in the long lineup of the company's numbered designs, is burning 70 percent less fuel per unit of distance than the Boeing 707 first flown in 1957, and the trend line shows average fuel burn declining by about 2 percent a year since 1960.[85]

That, and larger plane sizes, has made flying considerably cheaper, allowing billions of people to fly: since 1960 the global total of revenue passenger-kilometers has risen exponentially (with only a few years of stagnation or decline, caused by the 9/11 attacks in 2001, the economic crisis of 2008–2009, and the Covid-19 pandemic of 2020–2022), from less than 100 billion to 8.68 trillion in 2019, more than a 90-fold increase in six decades.[86]

And so the point has been made, albeit not without a small flood of numbers and graphs. In order to revolutionize transportation (with internal combustion engines on land, on water, and in the air), vastly increase industrial production (with electric motors doing everything

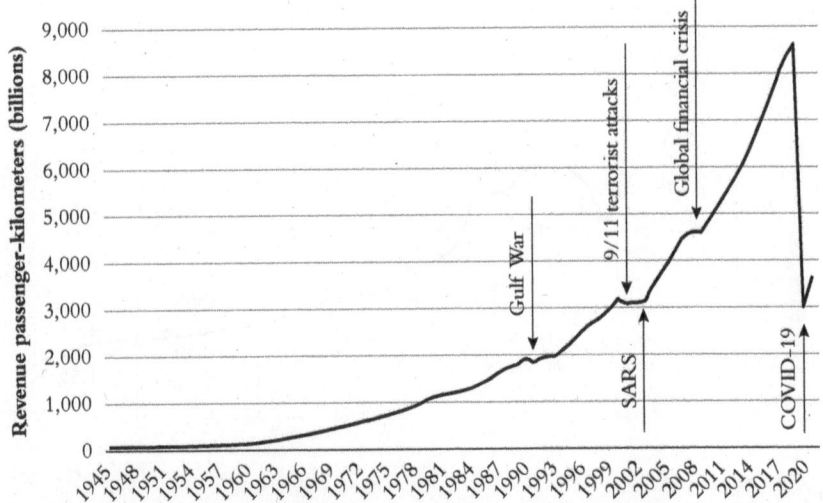

Only Covid-19 temporarily ended the decades-long exponential rise of revenue passenger-kilometers (previous downturns were small and short-lived).

from heavy lifting to precision drilling and metal machining), and meet the rapidly rising demand for electricity, we have not needed (some exceptions aside) faster-rotating diesel engines than those developed before 1900, faster-rotating steam turbines than the first pre-1900 commercial designs, faster-rotating motors than those conceived during the 1880s, or faster-rotating gas turbines than those conceived during the 1930s.

The rotations of those inventions and the early commercial designs sufficed for nearly all later needs, and what made the difference was not higher rpms but more powerful, more efficient, and more reliable machines. To rephrase, higher speed alone would not have led to the eventual mass-scale adoption of those machines resulting in unprecedented levels and rates of economic growth. All of them required subsequent decades of technical development—and that, in turn, was predicated on many scientific, technical, and managerial advances, including new material production and new ways of organizing production.

New materials that made those higher efficiencies, lower weights, and longer service spans possible have ranged from supplies of chromium, nickel, and molybdenum—metals needed for superior steel alloys (that quest had to begin with better methods of geo-physical prospecting)—to the synthesis of scores of new plastics and the creation of new composite compounds that now make up most of the mass of commercial jetliners. Organizational steps have ranged from Toyota's famous Kaizen—continuous improvement that seeks assiduous attention to detail and quality rather than aiming at maximized speed—to the outsourcing of products and tasks: there is now no equivalent of Ford's famous River Rouge plant in Detroit, with raw materials coming in and cars rolling out, as all cars are assembled from parts sourced from many suppliers on different continents.[87]

The speed of any process must eventually reach its limits, be it because of social, safety, and environmental concerns (the posted maxima of road travel), economic considerations (the speed of commercial air travel, the cost of fast container shipping), or fundamental physical constraints (the hardness of materials used in machining, maximum pressures, and the temperatures of specific operations). Epochal speed jumps introduced by fossil-fueled converters (engines and turbines) and by electric motors made all tasks—from food production and construction to manufacturing and the transportation of people and goods—considerably faster than during the premodern era, with most gains being not just single-digit multiples but (almost magical) jumps of several orders of magnitude, with the speed of computation and communication topping the scale.

Practical inventions of the 1880s and 1890s (internal combustion engines, steam turbogenerators, electric motors) and the introduction of gas turbines during the late 1930s were not driven solely or primarily by higher rotary speeds attainable by these machines. As I have documented in this chapter, these inventions have mostly retained (or only marginally increased) the typical speeds of their initial commercial applications. Consequently, it is both a misinterpretation and an oversimplification to conclude as Beaudreau does that "by the late 1960s/early 1970s, most machines in the manufacturing, mining and transportation sectors had reached their maximum speed. The evidence is there for all to see. Automobile, airplane, power tool, household appliance (washers, dryers) speeds have not increased since the 1960s"—and then to claim that this brought "to an end a century of speed-up-based productivity growth."[88]

The 20th-century development of rotating machines had far more (and in most cases nearly everything) to do with their efficiency, mass/power ratio, cost, and reliability than with increasing their original speeds. Perhaps most notably, the world could not have been transformed by intercontinental near-sonic flight if the developments of jetliners had not gone beyond the Boeing 707 launched in 1957 (its designed cruising speed was a bit higher than that of the Boeing 787, the company's latest and much more efficient wide-body model), or if the US Airline Deregulation Act of 1978 had not ended government-imposed restrictions on routes and fares and led to the unprecedented competition that has resulted in lower prices, better planes, and safer flying (but also in more congestion and delays).[89]

In those cases where speeds have shown a long history of increases (with the speed of cutting tools being perhaps the most outstanding example), it is, of course, inevitable that the gains will eventually diminish and cease, and that further productivity gains will be realized only by deploying more machines (that is, by increasing capital investment). Such speed plateaus have been a factor in the (mostly post-1970) global productivity decline—but not the only, or even most prominent, one. The modern productivity slowdown, noticeable in all advanced economies, has been repeatedly examined by economists, with (predictably) no solid agreement as to its principal causes or remedies.

Robert Gordon, an American economist who spent decades studying the history of US productivity, identified a slowdown in the overall innovation process as the leading factor, while other economists have pointed to the slowdown of global economic growth, the lack of effective demand, and increasingly debt-driven development. Meanwhile a 2021 study attributed the slowdown to a combination of the increasing mismeasurement of output, lower rates of growth of capital per worker, a slowdown in global trade, and changes in capital allocation.[90] In any case, it is most unlikely that what is by now a protracted (more than half a century) phenomenon affecting increasingly complex global economic development has (even when

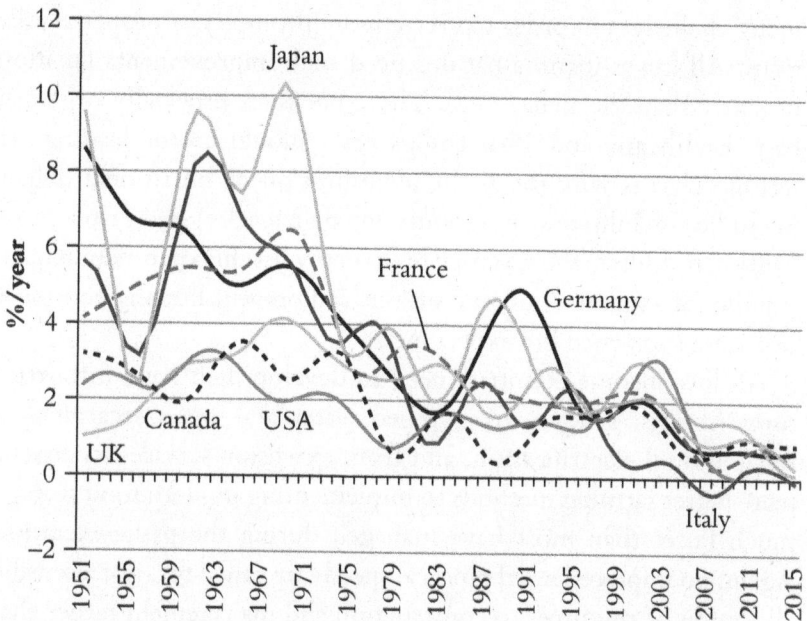

The long-term downward trends of labor productivity in G7 countries.

the historic trends showed such a clear "end of speed") a simple explanation in the change (or stagnation) of a single physical variable.

All too obviously, the global civilization of the early 21st century has many more, and more fundamental, problems than running out of opportunities to increase rotary or linear speeds. They range from malnutrition (still affecting nearly a billion people, or every eighth inhabitant) and increasing economic inequalities, to many environmental changes and degradations (from plastic pollution to warmer oceans), to existential uncertainties (be they prolonged wars or pandemics). Redressing, managing, and averting these problems is not predicated on any mechanical speed-ups: we have the means to do better, but we lack the commitment to adopt different priorities and pursue them with conviction.

Of course, there are still many instances where higher speeds would be helpful, but I would argue that these needs are almost solely in the category of *sensu lato* speeds and require no technical advances or machine redesigns but rather the better management of many challenges in order to arrive more promptly at proper resolutions. All lower-income nations need such improvements far more urgently than the richer ones. Those needs are especially urgent in basic healthcare, and "first things first" should be the leading criterion. That is why the faster identification of nutritional deficits (including debilitating micronutrient deficiencies) and faster intervention to address them is very high on my list: inexpensive adequate supplies of vitamin A or iron or zinc can prevent lifelong complications and shortened life expectancies.[91]

All low-income countries need to develop their basic infrastructures (the list is long, from piped water and water treatment to reliable rural electrification, and from extension services to disseminate better farming methods to implementing modern food storage) much faster than most have managed during the past generation. Again, no unprecedented physical speeds are called for, just a sensible allocation of resources to construction and management rather than destruction and violence—the near-chronic condition in more than a dozen African nations whose infrastructures (from roads to schools) are among the world's least adequate.[92]

But the rich world will never let go of its speed dreams, the quest for moving physically faster is here to stay, and new resources keep going into developing not just magnetically levitated super-rapid trains and new designs of supersonic airplanes, but even rockets with the eventual intent to use them in intercity transport. Those who think about rationalizing sensible priorities might feel exasperated, but we live in the age of nearly unlimited high-tech hype, where products of overactive imaginations are constantly misrepresented by the media (ever eager to push new "content") as soon-to-come quotidian realities.

Maglev (magnetic levitation, with magnets suspending and propelling a train) is the oldest of these aspirations, with the first concepts going back to the early 20th century and the first experiments starting in Japan and Germany during the 1980s.[93] So far, the only notable commercial project is the German-designed Pudong–Shanghai line (since 2004), and there are also three shorter links in South Korea and China. Construction of the Tokyo–Osaka maglev line has seen repeated postponements, and the latest target for opening the first section is now the late 2020s. In 2013, Elon Musk published his proposal for the Hyperloop Alpha, and this latest revival of a surprisingly old idea for very rapid train travel was greeted by those unfamiliar with the long history of such proposals (going back to the early 19th century!) as the beginning of imminent advances.[94] But many endorsements and announcements of the accelerated development of new intercity projects (in the US, Europe, and Asia) have not resulted in any new lines, and not even in any new firm targets. But that makes no difference—there will always be new claims about a soon-to-be launched maglev somewhere.

And it did not take long after Concorde's demise for some companies, and even some governments, to begin thinking again about supersonic transport (SST). Russia's Central Aerohydrodynamic Institute is designing a Mach 1.6 plane for 60–80 passengers, and so is the Tupolev Design Bureau that produced a copy of the Concorde during the 1970s. By 2019, four American companies— Aerion Supersonic, Spike Aerospace, Lockheed Martin, and Boom Technology—were developing supersonic planes.[95] Aerion Supersonic folded in 2021, Spike Aerospace promised in 2019 to deliver its first plane in 2023 but has nothing to offer, and Lockheed's plans depend on a new engine with no set date in sight.

That leaves Boom Supersonic and its Mach 1.7 (formerly Mach 2.2) Overture for 55 people. In 2021, construction of a new factory was to start the following year; in 2023 the company was to begin construction in 2024, with planes in service by 2029: that would be faster than Boeing's certification process for the 787. Moreover (as already noted), the company's CEO set the goal of flying "anywhere in the world in four hours for 100 bucks."[96] Of course, reality will eventually prevail: SST is not an inevitable next step in global aviation.

But could we not go beyond SST and use rocket-powered transportation on Earth? In 2017, Elon Musk's SpaceX made such a proposal, with a video depicting passengers taking:

> a large boat from a dock in New York City to a floating launchpad out in the water. There, they board the same rocket that Musk wants

to use to send humans to Mars by 2024. But instead of heading off to another planet once they leave the Earth's atmosphere, the ship separates and breaks off toward another city—Shanghai. Just 39 minutes and some 7,000 miles later, the ship reenters the atmosphere and touches down on another floating pad . . . Other routes proposed in the video include Hong Kong to Singapore in 22 minutes, London to Dubai or New York in 29 minutes, and Los Angeles to Toronto in 24 minutes.[97]

Musk also claimed that this rocket transport would begin flying within five years (that was in 2022) and that the cost would be "about the same as full fare economy in an aircraft."[98] Promoting a project with an utter disregard of its practicality, cost, and safety requires an extraordinary degree of hype; taking it seriously requires an equally extraordinary degree of uninformed credibility that is apparently never in short supply among the endlessly gullible mass media ("just might work," as NBC put it). There was no rocket taking humans to Mars in 2024, and readers who expect an intercontinental rocket taking people from New York to Shanghai will need both extraordinary patience and uncommon longevity. Many dreams of higher speed will remain just that—and for decades to come!

Coda

I could have ended the book on that cautious note, but I think I owe my readers a coda, a short addition addressing three facts. First, that my skeptical perspective regarding the boundless future of speed is not the prevalent take on the topic during the third decade of the 21st century, and that far greater advances are widely anticipated. Second, that the efforts aimed at (no matter how well-intentioned) deliberate speed restrictions should not be guided by incorrect analyses and by simplistic convictions. Third, that speed in modern societies has been embraced in ways that are as prominent as they are questionable and often even irrational, and this reality makes any changes more difficult.

The first sentiment, so different from my skepticism, has been well summarized by the already cited 2018 Rand Corporation study about speed and security. It concludes "there is a profound sense that life is speeding up . . . the world finds itself in a transitional phase to greater speed . . . the pace of almost everything is seemingly zooming beyond the technology it is replacing . . . technological developments and social dynamics are working in tandem to shift society into hyperdrive . . . A central thread is the hypothesis that this iteration of acceleration will be more intense than prior ones."[1]

And the "almost everything" designation is made more explicit by a list of those expected accelerated developments that they use to make scenarios not for 2100 but for 2040! Their "host of domains" includes "production, movement and transportation, communication, high-frequency trading and financial transactions, cognitive processing, data-sharing, information and knowledge transfer, weapon (e.g., missile) deployment, natural selection and evolution, computational processing, collective social organization, and cultural change."[2]

Moreover, Robert Colvile (of the *Daily Telegraph*) assures us (in *The Great Acceleration: How the World is Getting Faster, Faster*) that the positives of all-around acceleration outnumber the negatives, that further acceleration is truly inevitable, and that we should rush to embrace it![3] Of course, at this point, careful readers of this book have seen enough of my references to fundamental biophysical realities to be ready to join me in asking some obvious questions regarding many basic functions of modern civilization: how will they be part of that "greater speed," how will that shift be "more intense than prior ones," and should we really all "rush to embrace it"?

As for the first three categories listed in the Rand Corporation study—production, movement and transportation, and communication—there are many fundamental questions to ask. About food: Will we be able to harvest our crops in a small fraction of the time it takes us now? Will we be able to eliminate cattle and eat only cultured beef? Is there a 10-day wheat or two-week steak coming? About materials: Will we accelerate today's stepwise processes used to produce steel or ammonia to last mere minutes? For construction: Will we shorten the average time needed to build (or 3-D print!) high-rises to just days or weeks? About transportation: Will we move, say, 200,000 (instead of 20,000) steel containers on a single mega-ship from Shanghai to Hamburg in five instead of 50 days? Will we have noiseless SSTs flying everywhere and at low cost? About information: How can we transfer bits faster than at the speed of light/electricity? Are there no fundamental limits on computation?[4] Does our current high-frequency trading really need even faster speeds?

This list of questions could be greatly extended, but even these few (none of them eliciting enthusiastically positive replies) make it clear that the category of "almost everything" going into "hyperdrive" by the mid-21st century would not appear to be all that extensive. But the Rand study does not shy away from imagining the results of these anticipated accelerations as realities by 2040: from genetic plastic surgeries (to enhance everything from athletic abilities to fluency in languages) to brain hacking, from 3-D (and 4-D) printers not only "capable of printing and auto-assembling virtually anything" but doing that at home, to distant cities reached by "hyperloops" in a matter of minutes. I am always skeptical about any specific long-range forecasts (and about the now-so-prevalent techno-hype in particular), and I always remind my readers that no scenario, and no collation of expert opinions (a category that is not so easily defined in the electronic age), can tell us where we will be decades from now, and what consequences we will have to face.

At the same time, I concede, readily, that even if there will be no networks of maglev trains, no iBrain implants, no cheap intercity (or interplanetary) rockets, and (despite what we were repeatedly told in 2023) little evidence of an all-encompassing AI superiority by the year 2040, there is no doubt that there are some encouraging prospects for developments where higher speeds (in N/T or higher-frequency categories) could bring many significant benefits (above all in both preventive and curative healthcare, speeding up diagnoses and treatments)—as well as some worrisome concerns about changes coming too fast for us to cope with their (still poorly understood) consequences (including worries about AI's reach and about security, and the heightened probability of deadly conflicts fought by automated systems).

And there is yet another option instead of simply waiting for that allegedly inevitable new round of accelerations to overwhelm us (*pace* the Rand scenario) or working assiduously to make it happen even faster. Should we not try to bring about more balanced outcomes? Should we not pursue the option that deliberately combines some still much-needed accelerations with some desirable decelerations? This possibility was not as realistic to contemplate when the modern high-speed society began taking off during the 19th century, but by now the accumulation of our advances—as well as the drawbacks of excessive speed—makes it worth considering. This means going beyond the simple contrarian perceptions concerning the genesis of the modern devotion to speed and its eventual impacts.

The modern preference for speed (some might call it an obsession with it, or even the tyranny of) has not come about simply as an inevitable extension of an innate human propensity that was eventually translated by a largely organic process of technical advances into quotidian realities—a quest that has led to unprecedented combined gains of energy, mass, and speed that have rewarded billions of people with benefits unavailable even to the richest rulers of the early modern era (from avoiding death in infancy to sleeping in a warm room in winter). But nor has it been an unnatural outcome of a grand strategy and of a deliberately pursued quest motivated solely by profit and power, realized for the benefit of ruling minorities in total disregard of noise, pollution, safety, health, and dignified living, and leading to large-scale immiseration, urban squalor, and the rising inequality of incomes.

These polar stances have been formulated and defended by historians, economists, and politicians for more than 150 years.[5] But modern speed-driven transformations cannot be explained by forcing them into a simple framework. Human achievements never come without their downsides, and those unwelcome consequences tend to have highly uneven distributions. Higher speeds and their effects have been no exception. Speed brings perils and succor: excessive speed, in so many guises, kills—but receiving the fastest possible post-event care is the key to recovery from strokes or preventing death from internal bleeding, and the Covid pandemic showed, once again, the benefits of speedy mass-scale vaccination (particularly for older people).[6] The quest for speed causes inequalities, squalor, noise, and pollution (too many modern cities are witness to that)—but it also creates unprecedented economic opportunities, prolongs our lives, saves time, and makes the quotidian life easier (from long-distance travel to near-instant interior heating and air conditioning).[7]

But it is not difficult to make a definite case against the commonly held myth of speed as an unqualified benefit, and this book offers solid arguments in favor of limited speeds (both the L/T and N/T versions) and of deliberately imposed restrictions. How far this sentiment can go beyond the legal limits that are strongly supported by risk evaluation and by the studies demonstrating which speeds are compatible with the minimization of injuries and fatalities (be it on the roads or in workplaces) is uncertain. Advocacy for and the practice of deliberately restricted "life speeds" has been a reaction to a long list of undesirable conditions and outcomes associated with speed: social critics have charged speed with everything from the homogenization of lives and environments to depriving people of variety and texture in their lives; from causing stress, alienation, and exhaustion, to the loss of agency, feelings of powerlessness, the collapse of work/life distinctions, and mental and physical ailments.[8]

These undesirable states are to be remedied by supervised therapies or regular practices in search of decompression, calm, and tranquility based on stillness and slowness: sitting without distractions, listening to slow music, deep breathing, interacting with friendly animals, staying in rural respite facilities—all in the quest for stillness as a productive state.[9] This can be done in settings ranging from expensive facilities catering to celebrities to a quiet clearing in the woods or an isolated lakeshore. Undoubtedly, more people are seeking these experiences (some looking for stillness in movement)—but their total must be orders of magnitude smaller than the nearly 9 billion passengers who boarded commercial airplanes in 2023.[10]

Similar disparities apply when contrasting dominant experiences with attempts to promote slow cities and slow food. The Slow Food movement started in Italy in 1986 (after a demonstration at the proposed site of a McDonald's at Rome's Spanish Steps), the *Slow Food Manifesto* was published in Paris three years later, the Austrian Society for Deceleration of Time was founded in 1990, and since 1999 Norway has hosted the World Institute of Slowness.[11] Such steps have their undoubted appeal, but they are also tinged with elitism and wishful thinking. The Slow Food movement has not slowed down the global march of McDonald's. Since the early 1990s the company's annual profit has quadrupled: it has updated its menus and added franchises in such large markets as China, India, Pakistan, Vietnam, Egypt, and South Africa, and it now operates in more than 100 countries.[12]

The extreme form of speed abhorrence is to prefer walking to cars. Famously, Ivan Illich, an Austrian Catholic priest and a famous social critic, made that case in 1974 in his book on energy and equity: "People on their feet are more or less equal," but putting "more than a certain horsepower behind any one passenger . . . has reduced equality among men, restricted their mobility to a system of industrially defined routes and created time scarcity of unprecedented severity."[13] And after he accounted for the time a typical American needs to spend earning money in order to buy (or lease) a vehicle, as well as insuring, fueling, and repairing it, he concluded that "the model American puts in 1,600 hours to get 7,500 miles: less than five miles per hour. In countries deprived of a transportation industry, people manage to do the same, walking wherever they want to go . . ."[14]

Attempts at total cost assessments should always be welcome, as they offer a more realistic appraisal of overall burdens—but this denigration of driving and car speed is based on questionable foundations. Illich did not explain how he arrived at his totals, but I used the 2022–2023 US annual averages (for driving time, distance driven, after-tax income, and costs of a new vehicle, gasoline, insurance, and maintenance) to find that a typical American car owner spent 293 hours driving (an average speed of 74 km/h) and another 333 hours earning monies for annual vehicle use, resulting in an adjusted speed of about 35 km/h rather than the 7.5 km/h claimed by Illich.[15]

Further adjustments result from taking into account the cost of traffic accidents, and we do not need to make any elaborate estimates: the National Highway Traffic Safety Administration put the economic and societal impact of motor vehicle crashes in 2019 (the last year unaffected by Covid restrictions) at $340 billion.[16] With cars involved in just over half of all road accidents, that works out to about $700 per licensed driver, adding about 25 hours of additional earnings to the adjusted cost, and lowering the average speed to 33.3 km/h. This value must be compared with the similarly adjusted rate for walking: its energy cost is not (as implied by Illich) free, as an hour of moderately fast walking claims an additional 300 kilocalories of energy.

Every preindustrial society dependent on subsistence agriculture would have required at least 1 hour of additional agricultural labor to secure that food intake (as opposed to resting), halving the adjusted speed of an hour-long walk to less than 4 km/h—and making driving still nearly 10 times faster than walking, a gain that obviously holds a planet-wide appeal. An order of magnitude is clearly an irresistible speed difference that has helped to induce and then maintain high rates of North American car ownership, and that makes it clear that preconceived moralizing is no substitute for factual analysis. Of course, calculations for the most congested American cities (Chicago,

Boston, New York) would result in less impressive speed gains, and the full cost pricing of car ownership would yield even lower speeds in most low-income countries. We now have a thorough analysis of traffic in 1,200 cities in 152 countries that shows American urban speeds to be the world's highest, commonly 40–50 percent faster than in Europe and more than double the averages in the worst-off cities in India, Pakistan, or Bangladesh.[17]

But this has not stopped the car's universal appeal, and no recent development demonstrates it better than the rapid rise of private car ownership in China. In the early 1980s, soon after Deng Xiaoping's economic reforms began, the country had only about 1 million cars, by 2022 it sold domestically more than 26 million units a year, and in 2023 it became the world's largest exporter: decades of Maoist exhortations to lead frugal lives (made redundant by the economy's barely subsistence nature) and the need to reckon with some of the world's highest urban densities have made no difference to the car's rapid conquest of post-Mao China.[18]

To explain the car's global appeal, we must add status to speed. Nobody has offered a better explanation of this fundamental reality than Kenneth Boulding, an American economist and Illich's contemporary. He knew that a car "turns its driver into a knight with the mobility of the aristocrat . . . The pedestrian and the person who rides public transportation are, by comparison, peasants looking up with almost inevitable envy at the knights riding by in their mechanical steeds. Once having tasted the delights of a society in which almost everyone can be a knight, it is hard to go back to being peasants."[19]

Of course, in the long run, circumstances and incentives will change (there are already some signs of retreat from universal car ownership) but the advantages of speed and social elevation mean that this machine's prominent place in modern society is not going to disappear soon.

This place is even more remarkable given the already noted (often mortal) perils of driving, but we also have a convincing explanation for this exceptional risk tolerance. Chauncey Starr, an American engineer and scientist, showed that if people think they are in control (as they think they are while driving), they will tolerate risks up to three orders of magnitude higher than when they don't feel in control.[20] The last pandemic showed this contrast, once again, to perfection, as so many people were horrified by the idea of mandatory vaccination, imputing to it exceptionally high risks while having no fear of speeding.[21] Alternatively, familiarity (repetitive activity) has been proposed as the main reason for tolerating some demonstrably high risks, and the combination of the two factors may be the best explanation.[22]

And Illich, himself a very frequent flyer, was also quite wrong about the potential of global aviation. In 1974 he noted that "barely 0.2 per cent of the entire US population can engage in self-chosen air travel more than once a year"—but by 2019 (before Covid affected the industry), airlines handled 926 million domestic and international passengers (nearly three times the country's population total) in the US and 4.5 billion passengers worldwide (nearly 60 percent of the global population total).[23] For Illich, "the occasional chance to spend a few hours strapped in a high-powered seat" made a traveler "an accomplice in the distortion of human space"—but billions of people are now submitting to those distortions eagerly and repeatedly, and all forecasts indicate a further large-scale expansion of global flying.[24]

And while hundreds of millions of people are now so well-off that they can choose to live a quieter, cleaner, and "limited-speed" middle- and upper-income existence working in climatized offices or in the comfort of their homes, such choices are unavailable to the hundreds of millions who are trying to move up the well-being ladder and are forced to endure the relentless demands of jobs where dangerous speed is unavoidable. Workers in meatpacking plants, whose risky work I described earlier in this book, are but one of many categories of such exertions: as I have also noted, the rapid "fulfilment" of Internet merchandise orders—picking the items from shelves in giant warehouses, packaging and dispatching the envelopes and boxes—has become another large-scale business where speed endangers human well-being.

Moreover, many of these jobs provide essential food and material inputs for the world's affluent economies. Growing vegetables for the EU under plastic tents (with temperatures reaching up to

50°C) in southern Spain (with covered areas visible from space) is an example of the first category; the primitive and dangerous hand-mining of cobalt (now indispensable for car batteries) in Congo of the second.[25] The latest addition to these dangerous, speed-demanding jobs is the still-growing category of delivery services. Food delivery drivers working for major companies (in the US, DoorDash and ChowNow; in China, Meituan and Ele.me) or for small independent outfits in the world's largest cities are particularly vulnerable, especially when using motorcycles.[26]

More nuanced understanding and unprejudiced evaluations of the modern quest for speed are needed if we are to provide more balanced guidance for what is to come. That must begin with more incisive looks at the role of speed, at its benefits and its undesirable consequences, and at the need for further increases as well as formulating and enforcing its limits. Commercial flying is one of the best examples of the need for this balanced approach. We have machines capable of flying safely over continents at near-sonic speeds, but, too commonly, our airports are sites of failed organization, frequent long delays, cancellations, and seemingly endless waiting—to say nothing of the fact that, although we know better, we have yet to adopt more rational boarding procedures that are speedier than the currently used sequences.[27]

Finally, for better or worse, we must reckon with speed as an expression and a marker of the modern age. In the Preface I noticed the remarkable absence of speed both in classic writings and in many modern comprehensive discourses on economics, energy, and the environment, disciplines where such remarks could be expected. Was it just a coincidence that it was a poet who first captured the ethos of a new era at its very beginning? In 1819, just as mobile steam power began breaking the long reign of natural speed limits, Lord Byron, in his *Don Juan*, maintained that:

> Now there is nothing gives a man such spirits,
> Leavening his blood as cayenne doth a curry,
> As going at full speed—no matter where its
> Direction be, so 't is but in a hurry.[28]

In any case, that presaged the energetic and mechanical accomplishments of the 19th century, when increased speed—of travel, of grain harvesting, of metal smelting, of manufacturing, of communicating—became a sign of modernity. And when, almost a century after Byron wrote those lines, the English poet William Ernest Henley published *A Song of Speed*, that long poem captured the spirit of the age, seeing:

> Speed as a rapture:
> An integral element
> In the new scheme of Life

More than that, in grand Victorian fashion he linked God and a fast machine:

> And at times, when He feels
> That His creatures are doing
> Their best to assert
> Their part in His dream,
> He loosens His fist
> And a miracle slips from it
>
> . . .
>
> This amazing Mercédes,
> With Speed—
> *Speed in the Fear of the Lord.*[29]

As a poem, you may like it, tolerate it, or dismiss it—but the admiration of a fast machine is genuine.[30] Less than a decade later, Giacomo Balla, an Italian Futurist painter, tried to convey a similar emotion on a canvas when he attempted to capture the essence of a speeding car.

In any case, that appeal has always had universal reach: half a century later, another man smitten by another vehicle came from the opposite end of the political spectrum. The introduction of the Citroën DS 19 at the Paris Motor Show in October 1955 (with DS pronounced as *déesse*, "goddess") moved Roland Barthes, a famous French public intellectual and self-labeled Marxist, to write, not unlike Henley, "that cars today are almost the exact equivalent of great Gothic cathedrals: I mean the supreme creations of an era . . .

Giacomo Balla: *Speeding Automobile* (1913).

consumed in image if not in usage by a whole population which appropriates them as purely magical objects . . . It is obvious that the new Citroën has fallen from the sky . . . Speed here is expressed . . . as if it were evolving from a primitive to a classical form."[31]

I must admit that I, too, was taken by that design when I first saw Citroëns belonging to the French embassy driven around Prague in the early 1960s, when I was studying at Carolinum University. But for me that car served mostly as a sleek messenger of what was

The Gallic embodiment of speed—the Citroën DS 19.

possible on the other side of the Iron Curtain, rather than an enviable embodiment of speed. Of course, such reactions are highly personal: I am not an automotive afficionado. More than that, unless it is a hunting cheetah, I do not particularly admire terrestrial speeds; my real awe at natural speeds is reserved above all for hurricanes and volcanic eruptions, and as far as anthropogenic speeds go, I have always appreciated great airplane designs.

But despite my utilitarian attitude—I see cars as no more than metallic machines on rubber wheels used to get people from A to B—I did, eventually, embrace driving and (within limits) automotive speed. I lived without a car, in Europe and then in the US, for the first 30 years of my life, but then a succession of Hondas (Accords first, then after 20 years the switch to Civics, now more massive than our first Accord) became indispensable, and it is very hard for me to think about a time when I will have to give up driving. Convenience, choice, and feelings of independence and freedom might be my main justifications, but how to disentangle them from speed? I have never had, or caused, an accident in my life, but I cannot say that I have never driven faster than various posted speed limits!

Car appeal has yet to run its global course, but more than a hundred years after Henley's song of speed, and nearly a lifetime after Barthes's unlikely veneration of another fast vehicle, we should know better than to become infatuated with metallic symbols of land speed. The understanding we gained during the 20th century demands that we should be much less impressed by speed, and avoid conflating future advances with its increases. This means that we should manage most of speed's manifestations much better during the 21st century, and that we should not continue doing things with so much disregard for the environment and social equity as we have while creating, so speedily, the modern world.

That world was built on the scientific and technical breakthroughs of the long 19th century (1800–1913), and on their perfection and mass deployment (with some notable new additions) during the remainder of the 20th century.[32] Without these admirable advances we could not have, for the first time in history, lifted economies from millennia-long stagnation (or, at best, barely discernible and precarious improvement), assured adequate food supply for the entire planet (the remaining malnutrition is overwhelmingly due to failing access, not inadequate output), more than doubled average life expectancy, and provided a modicum of comfort for most of humanity, and a surfeit of goods and experiences for about a quarter of it.

This book has explained how higher speeds were fundamental in this quest, but that does not mean that further accelerations should be a major aim of our future endeavors. Obviously, we need to maintain speeds of many mobile and stationary machines that underpin modern economic performance, and we should even increase some of them, whenever we can combine technical or organizational improvements with affordable and rewarding applications. This should apply above all to speeds as higher frequencies of delivering items and services, with examples ranging from faster yet accurate diagnoses to speedier interventions to prevent the worst consequences of mental disorders.

But much more broadly, we need to adjust many speeds by considering a far broader range of concerns than those that contributed to their original development and adoption. That means focusing on those needs where higher speed—be it in strictly physical terms, or as the increased output of products or a higher frequency of interventions—improves quality of life and reduces income inequalities, and where lower speed (be it expressed as redesigned cities or food produced at a slower rate) does the same. The potential rewards of such actions are obvious given the pressures and problems besetting modern societies, from the perpetuation of poverty to the perils of illegal immigration, from the concerns about animal welfare to the overfishing of the oceans.

For many people, physical speed will always have that Byronic appeal of "nothing gives a man such spirits," but for the modern global civilization the main quality associated with speed should be the one expressed by the Latin proverb *non multa, sed multum* (not many, but much), its essence being that it is the quality rather than sheer quantity that matters. In a world of more than 8 billion people, large quantities, and hence considerable speeds, are unavoidable, and securing their smooth provision is essential for the functioning of a global civilization: modern society is too beholden to high speeds to revert to the pre-1800 mode of life.

But this indispensability does not preclude the quest for deploying, judging, and adjusting speed in the service of a higher quality of life, of lessening individual and collective risks, of moderating the impacts on the environment, of reducing inequalities, and of promoting greater social stability. Do we really need to be in a transitional phase to a yet greater speed? Should we work together to shift everything closer to the imaginary hyperdrive? Should our greatest goal be to create a society with even more intense accelerations? I would argue that from where we are today the pursuit of speed should not be a quest for further indiscriminate speed iterations, but rather a carefully considered combination of further gains, thoughtful limits, and justifiable retreats.

Acknowledgements

For the fifth time with the same Viking crew: thanks to Connor Brown, Natalie Wall and Gemma Wain.

References and Notes

Preface: From Speed's Absence to Singularity

1 National Library of Medicine. (2023). PubMed.gov (searches for: growth, speed). Accessed 2024.

2 Web of Science searches for growth and speed. Accessed via the University of Manitoba Libraries in December 2024.

3 Darwin, C. (1859). *On the Origin of Species*. London: John Murray; Thompson, D. W. (1917). *On Growth and Form*. Cambridge: Cambridge University Press; West, G. (2017). *Scale*. London: Viking.

4 Zipf, G. K. (1949). *Human Behavior and the Principle of Least Effort*. Cambridge, MA: Addison-Wesley Press; Mokyr, J. (2017). *A Culture of Growth*. Princeton, NJ: Princeton University Press.

5 Oleson, J. P., ed. (2008). *The Oxford Handbook of Engineering and Technology in the Classical World*. Oxford: Oxford University Press; Lécuyer, C. & Brock, D. C. (2010). *Makers of the Microchip*. Cambridge, MA: MIT Press; Odum, H. (1971). *Environment, Power, and Society*. New York: Wiley Interscience; Ayres, R. (2016). *Energy, Complexity and Wealth Maximization*. Berlin: Springer.

6 Feynman, R. (1989). *The Feynman Lectures on Physics: Commemorative Issue* (Vols. I–III). Redwood City, CA: Addison-Wesley Publishing Company. Feynman introduces speed in the eighth chapter of the first volume ("Motion") in subsections entitled "Speed," "Speed as a derivative," "Distance as an integral," and "Acceleration" (pp. 8-2–8-10 of the memorial edition). The citation is on p. 8-2.

7 Adams, H. (1904). A law of acceleration. In: *The Education of Henry Adams* (Chapter 34). New York: Houghton Mifflin. https://www.gutenberg.org/files/2044/2044-h/2044-h.htm.

8 Adams, H. (1919). *The Degradation of the Democratic Dogma*. New York: Macmillan Company, p. 308.

9 Smil, V. (2006). *Transforming the Twentieth Century*. New York: Oxford University Press.

10 Ulam, S. (1958). John von Neumann 1903–1957. *Bulletin of the American Mathematical Society*, 64(3.P2), 5.

11 Meyer, F. (1947). *l'Accélération evolutive*. Paris: Librairie des Sciences et des Arts; Halévy, D. (1948). *Essai sur l'accélération de l'histoire*. Paris: Self; Feynman, R. (2011). There's plenty of room at the bottom. *Resonance*, 16(9), 890–905; Moore, G. E. (1965). Cramming more components onto integrated circuits. *Electronics*, 38(8), 114–117; Moore, G. E. (1975). Progress in digital integrated electronics. *Technical Digest, IEEE International Electron Devices Meeting*, 11–13; Piel, G. (1972). *The Acceleration of History*. New York: Random House; Moravec, H. (1988). *Mind Children*. Cambridge, MA: Harvard University Press; Vinge, V. (1993). *The Coming Technological Singularity*. https://accelerating.org/articles/comingtechsingularity.html; Coren, R. L. (198). *The Evolutionary Trajectory: The Growth of Information in the History and Future of Earth*. Boca Raton, FL: CRC Press.

12 Kurzweil, R. (2001). The law of accelerating returns. https://www.karmak.org/archive/2003/01/art0134.html.

13 Kurzweil, R. (2005). *The Singularity Is Near*. New York: Penguin.

14 Wajcman, J. & Dodd, N., eds. (2016). *The Sociology of Speed: Digital, Organizational, and Social Temporalities*. Oxford: Oxford University Press; Steffen, W. et al. (2015). The trajectory of the Anthropocene: The Great Acceleration. *The Anthropocene Review*, 2, 81–98; McNeill, J. R. & Engelke, P. (2016). *The Great Acceleration: An Environmental History of the Anthropocene since 1945*. Cambridge, MA: Belknap Press.

15 Virilio, P. (2006). *Speed and Politics*. Los Angeles: Semiotext(e), p. 46.

16 Virilio (2006), p. 68.

17 Virilio (2006), p. 69.

18 Mumford, L. (1967). *The Myth of the Machine*. New York: Harcourt Brace Jovanovich, p. 294.

19 Virilio (2006), p. 78.

20 In Richard Burton's famous 1885 translation: "Whoever sitteth on this carpet and willeth in thought to be taken up and set down upon other site will, in the twinkling of an eye, be borne thither, be that place nearhand or distant many a day's journey and difficult to reach."

21 The passage from Homer's *Odyssey* reads: "She at once set herself to think how she could speed Ulysses on his way. So she gave him a great bronze axe that suited his hands . . . and then led the way to the far end of the island where the largest trees grew—alder, poplar and pine, that reached the sky—very dry and well seasoned, so as to sail light for him in the water."

22 Grimm, J. & Grimm, W. (1819). Der goldene Vogel (The Golden Bird). In: *Grimms' Fairy Tales.* https://www.gutenberg.org/files/2591/2591-h/2591-h.htm.

23 To cite the official source: "Hyperdrives allow starships to travel faster than the speed of light, crossing space through the alternate dimension of hyperspace. Large objects in normal space cast "mass shadows" in hyperspace, so hyperspace jumps must be precisely calculated to avoid collisions." https://www.starwars.com/databank/hyperdrive.

24 Andrews, G. et al. (2022). Speed and space: Rates of motion in health and wellbeing. *Wellbeing, Space and Society, 3,* 100112.

25 O'Hare, M. (2021, May 18). Boom Supersonic aims to fly "anywhere in the world in four hours for $100." CNN. https://www.cnn.com/travel/article/boom-supersonic-four-hours-100-bucks/index.html.

26 You can now easily check all those claims of fast Internet connections by simply clicking on https://fast.com/ which will show your download speed in Mbps.

Introduction—Universe of Speeds: Variables and measurements

1 Rogers, L. (2011, February 1). A brief history of time measurement. University of Cambridge Faculty of Mathematics. https://nrich.maths.org/articles/brief-history-time-measurement.

2 Payn, H. (1914). The well of Eratosthenes. *The Observatory, 37,* 287–288.

3 Walkup, N. (2010). Eratosthenes and the mystery of the stades: The basic problem. *Convergence.* https://old.maa.org/press/periodicals/convergence/eratosthenes-and-the-mystery-of-the-stades.

4 Stephenson, F. R. et al. (2016). Measurement of the Earth's rotation: 720 BC to AD 2015. *Proceedings of the Royal Society A, 472,* 20160404.

5 Jones, G. & Bikos, K. (2022, July 27). Earth is in a hurry in 2020. *Time and Date*. https://www.timeanddate.com/time/earth-faster-rotation. html.

6 Dunn, R. & Higgitt, R. (2014). *Finding Longitude*. London: Collins; Gould, R. T. (1923). *The Marine Chronometer: Its History and Development*. London: J. D. Potter.

7 Smeaton, W. A. (2000). The foundation of the metric system in France in the 1790s. *Platinum Metals Review*, *44*, 125–134; Bureau International des Poids et Mesures. (2018). *Resolution 1 of the 26th CGPM*.

8 Crosby, A. W. (1997). *The Measure of Reality*. Cambridge: Cambridge University Press.

9 Penzes, W. B. (2025). Time line for the definition of the meter. https://www.nist.gov/system/files/documents/pml/div683/museum-timeline.pdf.

10 Conférence générale des poids et mesures. (1889). *Comptes rendus des séances de la première Conférence générale des poids et mesures*. Paris: Gauthier Villars.

11 Bureau international des poids et mesures. (2025). *Le Système international d'unités (SI)*. Sèvres: BIPM.

12 National Institute of Standards and Technology. (2025). SI units—length. https://www.nist.gov/pml/owm/si-units-length.

13 Kaler, J. B. (1992). *Stars*. New York: Scientific American Library, p. 36.

14 For example, exoplanets Kepler-22b, Kepler-69c, and Kepler-62f are, respectively, 600, 2,700 and 1,200 light-years away: Harvey, A. & Howell, E. (2022). The 10 most Earth-like exoplanets. Space.com. https://www.space.com/30172-six-most-earth-like-alien-planets. html/.

15 Feynman (1989), pp. 8-2 and 8-3.

16 Shea, J. H. (1998). Ole Rømer, the speed of light, the apparent period of Io, the Doppler effect, and the dynamics of Earth and Jupiter. *American Journal of Physics*, *66*, 561–569.

17 Klinaku, S. (2013). Review of Michelson's and Fizeau's experiments. *Science Journal of Physics*, *2014*, sjp-102.

18 BIPM. (1983). Definition of the metre. https://www.bipm.org/en/committees/cg/cgpm/17-1983/resolution-1.

19 Average sailing times were affected by spells of no progress when becalmed, as well as by frequent mishaps including losing a mast and running aground.

20 Kessler, A. A. (1920). Animal-drawn transport in war. *The Military Engineer, 12*, 341–345.

21 Mercedes-Benz. (2025). Mercedes-Benz: Benz Patent Motor Car. https://www.mercedes-benz.com/en/innovation/milestones/benz-patent-motor-car/.

22 Gunston, B. (2002). *Aviation: The First 100 Years*. Hauppauge, NY: Barron's.

23 Smil, V. (2023). *Invention and Innovation*. Cambridge, MA: MIT Press.

24 Joyce, J. (1916). *A Portrait of the Artist as a Young Man*. New York: B. W. Huebsch, p. 87.

25 Wu, C. et al. (2017). An investigation of perceived vehicle speed from a driver's perspective. *PLoS ONE, 12*(10), e0185347.

26 Riggs, W. (2019). Perception of safety and cycling behaviour on varying street typologies: Opportunities for behavioural economics and design. *Transportation Research Procedia, 41*, 204–218; Colombet, F. et al. (2010). Impact of geometric field of view on speed perception. *Les collections de l'INRETS*, 69–79. http://dsc2015.tuebingen.mpg.de/Docs/DSC_Proceedings/2010/DSC10_07_Colombet.pdf.

27 Walton, A. et al. (2001). Drivers' biased perceptions of speed and safety campaign messages. *Accident Analysis and Prevention, 33*, 629–640.

28 Papić, Z. et al. (2020). Underestimation tendencies of vehicle speed by pedestrians when crossing unmarked roadway. *Accident Analysis & Prevention, 143*, 105586.

29 Thompson, P. (1982). Perceived rate of movement depends on contrast. *Vision Research, 22*, 377–380.

30 Sotiropoulos, G. (2014). Contrast dependency and prior expectations in human speed perception. *Vision Research, 97*, 16–23.

31 Royal Meteorological Society. (2025). The Beaufort wind scale. https://www.rmets.org/metmatters/beaufort-wind-scale.

32 Weight-for-length growth charts for infants and children can be seen at: Centers for Disease Control and Prevention. (2023). CDC growth charts. https://www.cdc.gov/growthcharts/cdc-growth-charts.htm.

33 Filling the gasoline tank of a passenger car (not of a gargantuan SUV) takes only about two minutes; raising the indoor temperature of a moderately sized house by 1°C should not take more than 15 minutes.

34 For the post-1960 record of China's annual GDP growth, see: World Bank. (2025). GDP growth (annual %)—China. https://data.world-bank.org/indicator/NY.GDP.MKTP.KD.ZG?locations=CN.

35 Oberlo. (2024). How many people have smartphones? https://www.oberlo.com/statistics/how-many-people-have-smartphones#.

36 Our World in Data. (2024). Coronavirus (COVID-19) vaccinations. https://ourworldindata.org/covid-vaccinations.

37 Coyle, D. (2014). *GDP: A Brief but Affectionate History*. Princeton, NJ: Princeton University Press.

38 The Grande Armée was in Moscow 83 days after crossing the Neman River, held the city for just 36 days, and after 57 days of retreat only some 15 percent of the initial force of 600,000 recrossed the Neman on December 14, 1812.

39 Central Intelligence Agency. (2025). About CIA in Afghanistan. https://www.cia.gov/legacy/museum/exhibit/on-the-front-lines-cia-in-afghanistan/.

40 Global Conflict Tracker. (2024). Instability in Afghanistan. https://www.cfr.org/global-conflict-tracker/conflict/war-afghanistan.

41 Dickinson, H. W. & Jenkins, R. (1927). *James Watt and the Steam Engine*. Oxford: Oxford University Press.

42 Smil, V. (2017). *Energy and Civilization*. Cambridge, MA: MIT Press, pp. 101–105.

43 And two US-made tractors are more than twice as powerful: Case IH Steiger 620 Quadtrac has 692 hp, New Holland T9.700 682 hp.

44 The P-51 Mustang had maximum speed of 710 km/h; modern jetliners cruise (now commonly for more than 12 and up to 17 hours) at about 900 km/h.

45 The UK, the first country to undergo industrialization, is now the best example of the opposite process: in 2021 only 9 percent of its GDP originated in manufacturing, compared to 19 percent in Germany and 27 percent in China: World Bank. (2023). Manufacturing, value added (% of GDP). https://data.worldbank.org/indicator/NV.IND.MANF.ZS.

46 Smil, V. (2021). *Grand Transitions: How the Modern World Was Made.* New York: Oxford University Press, pp. 25–28.

47 Smil (2021), pp. 3–7.

48 We3 Group. (2023). PET blow moulding machine. https://www. we3group.com/pet-blow-moulding-machine.html; American Society of Mechanical Engineers. (2023). Owens AR bottle machine. https:// www.asme.org/about-asme/engineering-history/landmarks/86-owens-ar-bottle-machine; Toyota. (2012). Production overview. https:// www.toyota-global.com/company/history_of_toyota/75years/data/ automotive_business/production/production/overview/index.html.

49 Intel. (2025). Announcing a New Era of Integrated Electronics. https:// www.intel.com/content/www/us/en/history/virtual-vault/articles/ the-intel-4004.html.

50 Strokes present the most acute challenge: a typical stroke patient loses nearly 2 million neurons or 1.8 days of healthy life for each minute of delay in treatment, and hence it is imperative to receive intravenous tPA (tissue-type plasminogen activator), the only medical therapy to improve outcomes for acute ischemic strokes, as soon as possible: Xian, Y. & Fonarow, G. C. (2018). The need for speed: Accelerating quality of care in acute ischemic stroke. *Circulation: Cardiovascular Quality and Outcomes, 11*(12), e005234.

51 Observatory of Economic Complexity. (2023). Economic complexity rankings. https://oec.world/en#.

52 Growth Lab. (2023). Ranking comparison tool. https://atlas.hks.harvard.edu/rankings.

53 Bejan, A. (2019). Why the days seem shorter as we get older. *European Review, 27*, 187–194; Di Lernia, D. et al. (2018). Feel the time. Time perception as a function of interoceptive processing. *Frontiers in Human Neuroscience, 12*, 74; Wittmann, M. (2009). The inner experience of time. *Philosophical Transactions of the Royal Society B, 364*, 1955–1967; Tien, J. M. & Burnes. J. P. (2000). On the perceived speed of time over time. *IEEE Xplore.* https://ieeexplore.ieee.org/document/885056; Wallach, M. A. & Green, L. R. (1961). On age and the subjective speed of time. *Journal of Gerontology, 16*, 71–74.

54 Bouskill, K. E. et al. (2018). *Speed and Security.* Santa Monica, CA: Rand Corporation, p. 1.

55 Kolmar, C. (2023). 25+ incredible US smartphone industry statistics. https://www.zippia.com/advice/us-smartphone-industry-statistics/.

56 Felton, N. (2008, February 10). Consumption spreads faster today. *The New York Times*. https://archive.nytimes.com/www.nytimes.com/imagepages/2008/02/10/opinion/100p.graphic.ready.html.

57 Asurion. (2025). The new normal: phone use is up nearly 4-fold since 2019. https://www.asurion.com/connect/news/tech-usage/; Wheelwright, T. (2022). 2022 cell phone usage statistics. Reviews.org. Other surveys showed averages between 144 and 205 a day. https://www.reviews.org/mobile/cell-phone-addiction.

58 Backlinko. (2024). Smartphone usage statistics for 2024. https://backlinko.com/smartphone-usage-statistics; Vecta Labs. (2023, July 25). Global mobile trends in 2023. https://vectalabs.com/global-mobile-phone-usage-trends-2023-a-vecta-labs-exploration/. But these figures can be just as easily cited as convincing proof of mass-scale deceleration: spending up to two and a half months a year on a mobile prevents, derails, and delays many other activities.

59 Elliott, C. (2019, July 20). Here are the disturbing reasons vacations are getting shorter. *Forbes*. https://www.forbes.com/sites/christopherelliott/2019/07/20/here-are-the-disturbing-reasons-vacations-are-getting-shorter/?sh=7cee0f1e61ad.

60 Bouskill et al. (2018).

1. Evolution: Non-intuitive time spans and speeds

1 Brans, P. (2022). Googol and googolplex. TechTarget. https://www.techtarget.com/whatis/definition/googol-and-googolplex.

2 Carnegie Science. (2025). Deep Carbon Observatory and Robert M. Hazen. https://hazen.carnegiescience.edu/research/deep-carbon-observatory. The volume of the Earth's deep biosphere is almost twice the volume of all oceans (2–2.3 billion km³).

3 Smil, V. (2008). *Energy in Nature and Society*. Cambridge, MA: MIT Press.

4 The slowest daily gains we can notice are the elongations of stems or opening of flowers of many plants—that is, speeds on the order of a few millimeters to a few centimeters a day.

5 Chambers, J. & Mitton, J. (2017). *From Dust to Life: The Origin and Evolution of Our Solar System*. Princeton, NJ: Princeton University Press.

6 Patterson, C. (1956). Age of meteorites and the earth. *Geochimica et Cosmochimica Acta, 10*, 230–237.

7 Canup, R. M. (2008). Accretion of the Earth. *Philosophical Transactions of the Royal Society A, 366*, 4061–4075; Yu, G. & Jacobsen, S. B. (2011). Fast accretion of the Earth with a late Moon-forming giant impact. *Proceedings of the National Academy of Sciences, 108*, 17604–17609.

8 Zhang, Y. (2002). The age and accretion of the Earth. *Earth-Science Reviews, 59*, 235–263.

9 Kleine, T. & Rudge, J. F. (2011). Chronometry of meteorites and the formation of the Earth and Moon. *Elements, 7*, 41–46.

10 Gradstein, F. et al., eds. (2012). *The Geologic Time Scale*. Amsterdam: Elsevier; The Geological Society of America. (2023). GSA geologic time scale. https://rock.geosociety.org/net/documents/gsa/time-scale/timescl.pdf.

11 Santosh, M. et al. (2017). Hadean Earth and primordial continents: The cradle of prebiotic life. *Geoscience Frontiers, 8*, 309–327.

12 Wu, J. et al. (2018). Origin of Earth's water: Chondritic inheritance plus nebular ingassing and storage of hydrogen in the core. *Journal of Geophysical Research: Planets, 123*, 2691–2712; Martin, R. G. & Livio, M. (2021). How much water was delivered from the asteroid belt to the Earth after its formation? *Monthly Notices of the Royal Astronomical Society: Letters, 506*, L6–L10.

13 Arrhenius, G. & Lepland, A. (2000). Accretion of Moon and Earth and the emergence of life. *Chemical Geology, 169*, 69–82.

14 Golding, S. D. & Glikson, M., eds. (2011). *Earliest Life on Earth: Habitats, Environments and Methods of Detection*. Dordrecht: Springer.

15 Wacey, D. (2012). Earliest evidence for life on Earth: An Australian perspective. *Australian Journal of Earth Sciences, 59*, 153–166; Allwood, A. C. et al. (2007). 3.43-billion-year-old stromatolite reef from the Pilbara Craton of Western Australia: Ecosystem-scale insights to early life on Earth. *Precambrian Research, 158*, 198–227.

16 Schopf, J. W. (2006). Fossil evidence of Archaean life. *Philosophical Transactions of the Royal Society B, 361*, 869–885.

17 Dodd, M. S. et al. (2017). Evidence for early life in Earth's oldest hydrothermal vent precipitates. *Nature, 543*, 60–65; Witze, A. (2017, September 27). Oldest traces of life on Earth may lurk in Canadian rocks. *Nature.* https://www.nature.com/articles/nature.2017.22685.

18 Luo, G. et al. (2016, May 13). Rapid oxygenation of Earth's atmosphere 2.33 billion years ago. *Science Advances, 2*, e1600134.

19 Schirrmeister, B. E. et al. (2011). The origin of multicellularity in cyanobacteria. *BMC Evolutionary Biology, 11*, 45.

20 Droser, M. L. & Gehling, J. G. (2015). The advent of animals: The view from the Ediacaran. *Proceedings of the National Academy of Sciences, 112*, 4865–4870.

21 Mángano, M. G. & Buatois, L. A., eds. (2016). *The Trace-Fossil Record of Major Evolutionary Events. Volume 1: Precambrian and Paleozoic.* Berlin: Springer.

22 Currie, P. & Padian, K. (1997). *Encyclopedia of Dinosaurs.* Amsterdam: Elsevier; Brusatte, S. (2012). *Dinosaur Paleobiology.* Chichester: Wiley.

23 van Holstein, L. & Foley, R. A. (2017). Hominin evolution. In: T. K. Shackelford and V. A. Weekes-Shackelford, eds. *Encyclopedia of Evolutionary Psychological Science.* Cham: Springer International Publishing AG, pp. 1–22.

24 Martínez, I. & Conde-Valverde, M. (2020). Mapping the ancestry of primates. *eLife, 9*, e55429; Stringer, C. (2016). The origin and evolution of Homo sapiens. *Philosophical Transactions of the Royal Society B, 371*, 20150237.

25 All of the following shares and 24-hour analogs are calculated from data in: The Geological Society of America. (2023). GSA geologic time scale. https://rock.geosociety.org/net/documents/gsa/time-scale/timescl.pdf.

26 Wegener, A. (1924). *The Origin of Continents and Oceans.* New York: E. P. Dutton.

27 Hess, H. H. (1962). History of ocean basins. In: A. E. J. Engel et al., eds. *Petrologic Studies: A Volume to Honor A. F. Buddington.* Boulder, CO: Geological Society of America, pp. 599–620.

28 Coats, R. R. (1962). Magma type and crustal structure in the Aleutian arc. In: G. A. Macdonald and H. Kuno, eds. *The Crust of the Pacific*

Basin. Washington, DC: American Geophysical Union Geophysical Monograph 6, pp. 92–109.

29 White, D. A. et al. (1970). Subduction. *Geological Society of America Bulletin, 81*, 3431–3432.

30 Vine, F. J. & Matthews, D. H. (1963). Magnetic anomalies over ocean ridges. *Nature, 199*, 947–949.

31 Wilson, J. T. (1963). Possible origin of the Hawaiian islands. *Canadian Journal of Physics, 41*, 863–870.

32 Wilson, J. T. (1965). A new class of faults and their bearing on continental drift. *Nature, 207*, 343–347.

33 SanAndreasFault.org. (2013). San Andreas Fault information. https://www.sanandreasfault.org/Information.html.

34 ETH. (2017). Introduction to tectonics. https://www.files.ethz.ch/structuralgeology/jpb/files/English/1Introtecto.pdf.

35 Dewey, J. et al. (1989). Tectonic evolution of the India/Eurasia collision zone. *Eclogae Geologicae Helvetiae, 82*, 717–734.

36 Meert, J. G. et al. (1993). A plate-tectonic speed limit? *Nature, 363*, 216–217; Zahirovic, S. et al. (2015). Tectonic speed limits from plate kinematic reconstructions. *Earth and Planetary Science Letters, 418*, 40–52.

37 Pusok, A. E. & Stegman, D. R. (2020). The convergence history of India-Eurasia records multiple subduction dynamics processes. *Scientific Advances, 6*, 8681.

38 Willett, S. D. et al. (2001). Uplift, shortening, and steady state topography in active mountain belts. *American Journal of Science, 301*, 455–485.

39 Jiang, W. et al. (2009). New model of Antarctic plate motion and its analysis. *Chinese Journal of Geophysics, 52*, 23–32.

40 NASA. (2009). SLR and GPS. https://cddis.nasa.gov/docs/2009/HTS_0910.pdf.

41 Willett et al. (2001).

42 Wang, C. et al. (1982). Dynamic uplift of the Himalaya. *Nature, 298*, 553–556.

43 Lavé, J. & Avouac, J. P. (2001). Fluvial incision and tectonic uplift across the Himalayas of central Nepal. *Journal of Geophysical Research, 106*, 26561–26591.

44 Tremblay, M. M. et al. (2015). Erosion in southern Tibet shut down at ~10 Ma due to enhanced rock uplift within the Himalaya. *Proceedings of the National Academy of Sciences*, 112, 12030–12035.

45 Hervé, F. & Ota, Y. (1993). Fast Holocene uplift rates at the Andes of Chiloé, southern Chile. *Revista Geológica de Chile*, 20, 15–23.

46 Nocquet, J.-M. et al. (2016). Present-day uplift of the western Alps. *Scientific Reports*, 6, 28404.

47 Pinter, N. & Brandon, M. T. (2005, July). How erosion builds mountains. *Scientific American*, 15(2), 74–81.

48 Fyon, A. (2020). Canada (Ontario) Beneath our feet. https://www.ontariobeneathourfeet.com/rising-land-isostatic-rebound.

49 The Tibetan plateau did not rise as a preformed entity, or solely because of crustal thickening driven by the India–Eurasia collision, but evolved gradually through tectonic compression and sediment infills: Spicer, R. A. et al. (2020). Why "the uplift of the Tibetan Plateau" is a myth. *National Science Review*, 8, nwaa091.

50 See, for example: https://www.thermexcel.com/english/program/pool.htm.

51 Hassan, A. et al. (2018). Evaluating evaporation rate from high Aswan Dam Reservoir using RS and GIS techniques. *The Egyptian Journal of Remote Sensing and Space Sciences*, 21, 285–293.

52 Denny, M. W. (1993). *Air and Water: The Biology and Physics of Life's Media*. Princeton, NJ: Princeton University Press.

53 World Rivers. (2020, March 28). How fast are rivers? http://worldrivers.net/2020/03/28/how-fast-are-rivers.

54 Bartsch-Winkler, S. & Lynch, D. K. (1988). *Catalog of Worldwide Tidal Bore Occurrences and Characteristics*. Washington, DC: US Geological Survey.

55 Laval, G. (2004). Ice streams—fast, and faster? Fleuves de glace: toujours plus vite? *Comptes Rendus Physique*, 723–734.

56 Sommer, C. et al. (2020, June 25). Rapid glacier retreat and downwasting throughout the European Alps in the early 21st century. *Nature Communications*, 11(1), 3209.

57 Di Mauro, B. et al. (2018). Saharan dust events in the European Alps: Role in snowmelt and geochemical characterization. *The Cryosphere*, 13(4), 1147–1165; Gutleben, M. et al. (2022). Wintertime Saharan dust

transport towards the Caribbean: An airborne lidar case study during EUREC⁴A. *Atmospheric Chemistry and Physics*, 22, 7319–7330.

58 Nolan, D. S. et al. (2014). On the limits of estimating the maximum wind speeds in hurricanes. *Monthly Weather Review*, *142*, 2814–2837.

59 Bierman, P. R. & Nichols, K. K. (2004). Rock to sediment-slope to sea with ¹⁰Be-rates of landscape change. *Annual Review of Earth and Planetary Sciences*, *32*, 215–255.

60 Lavé, J. & Avouac J. P. (2001). Fluvial incision and tectonic uplift across the Himalayas of central Nepal. *Journal of Geophysical Research*, *106*, 26561–26591.

61 Summerfield, M. A. & Hulton, N. J. (1994). Natural controls of fluvial denudation rates in major world drainage basins. *Journal of Geophysical Research*, *99*, 13871–13883.

62 Benaud, P. et al. (2020, July 15). National-scale geodata describe widespread accelerated soil erosion. *Geoderma*, *371*, 114378.

63 Thaler, E. A. et al. (2022). Rates of historical anthropogenic soil erosion in the Midwestern United States. *Earth's Future*, *10*, e2021EF002396.

64 Wuepper, D. et al. (2020). Countries and the global rate of soil erosion. *Nature Sustainability*, *3*, 51–55.

65 US Geological Survey. (2022). All earthquakes. https://www.usgs.gov/programs/earthquake-hazards/lists-maps-and-statistics.

66 National Science Foundation. (2005, June 1). Analysis of the Sumatra-Andaman earthquake reveals longest fault rupture ever. *Science Daily*. https://www.sciencedaily.com/releases/2005/05/050527104756.htm.

67 Socquet, A. et al. (2019). Evidence of supershear during the 2018 magnitude 7.5 Palu earthquake from space geodesy. *Nature Geoscience*, *12*, 192–199; Bao, H. et al. (2019). Early and persistent supershear rupture of the 2018 magnitude 7.5 Palu earthquake. *Nature Geoscience*, *12*, 200–205.

68 Tosi, P. et al. (2012). Earthquake sound perception. *Geophysical Research Letters*, *39*, L24301.

69 Shearer, P. M. (2010, June). Introduction to seismology: The wave equation and body waves. https://igppweb.ucsd.edu/~shearer/CIDER/notes.pdf.

70 Taylor, K. M. et al. (2011). Estimation of arrival times from seismic waves: A manifold-based approach. *Geophysical Journal International*, *185*, 435–452.

71 National Park Service. (2023). Volcanic explosivity index. https://www.nps.gov/subjects/volcanoes/volcanic-explosivity-index.htm.

72 Rowland, S. K. &. Walker, G. P. L. (1990). Pahoehoe and aa in Hawaii: Volumetric flow rate controls the lava structure. *Bulletin of Volcanology, 52,* 615–628; US Geological Survey. (1998, March 26). Volcano Watch—How fast does Hawaiian lava flow? https://www.usgs.gov/news/volcano-watch-how-fast-does-hawaiian-lava-flow.

73 Bercovici, D. & Michaut, C. (2010). Two-phase dynamics of volcanic eruptions: compaction, compression and the conditions for choking. *Geophysical Journal International, 182,* 843–864.

74 Carey, C. et al. (1996). Pyroclastic flows and surges over water: An example from the 1883 Krakatau eruption. *Bulletin of Volcanology, 57,* 493–511; Legros, F. & Kelfoun, K. (2000). On the ability of pyroclastic flows to scale topographic obstacles. *Journal of Volcanology and Geothermal Research, 98,* 235–241.

75 Gueugneau, V. et al. (2020). Dynamics and impacts of the May 8th, 1902 pyroclastic current at Mount Pelée (Martinique): New insights from numerical modeling. *Frontiers in Earth Science, 8,* 279; Rossano, S. et al. (1998). Computer simulations of pyroclastic flows on Somma–Vesuvius volcano. *Journal of Volcanology and Geothermal Research, 82,* 113–137; Gurioli, L. et al. (2005). Interaction of pyroclastic density currents with human settlements: Evidence from ancient Pompeii. *Geology, 33,* 441–444.

76 Wilson, C. J. N. & Walker, G. P. L. (1985). The Taupo eruption, New Zealand I. General aspects. *Philosophical Transactions of the Royal Society of London A, 314,* 199–228.

77 Dobran, F. et al. (1994). Assessing the pyroclastic flow hazard at Vesuvius. *Nature, 367,* 551–554.

78 Self, S. & Rampino, M. (1981). The 1883 eruption of Krakatau. *Nature, 294,* 699–704.

79 US Geological Survey. (2020). Ten ways Mount St. Helens changed our world. https://pubs.usgs.gov/fs/2020/3031/fs20203031.pdf.

80 Pierson, T. C. et al. (1990). Perturbation and melting of snow and ice by the 13 November 1985 eruption of Nevado del Ruiz, Colombia, and consequent mobilization, flow and deposition of lahars. *Journal of Volcanology and Geothermal Research, 41,* 17–66.

81 International Tsunami Information Center. (2025). How does tsunami energy travel across the ocean and how far can tsunamis waves reach? https://tsunami.ioc.unesco.org/en/pacific.

82 Lay, T. et al. (2005). The Great Sumatra-Andaman Earthquake of 26 December 2004. *Science, 308,* 1127–1133; Satake, K. (2014). The 2011 Tohoku, Japan, earthquake and tsunami. In: A. Ismail-Zadeh et al., eds. *Extreme Natural Hazards, Disaster Risks and Societal Implications.* Cambridge: Cambridge University Press, pp. 340–351.

83 Miller, D. J. (1960). *Giant Waves in Lituya Bay, Alaska.* Washington, DC: US Geological Survey.

84 Ward, S. N. & Day, S. (2001). Cumbre Vieja volcano—potential collapse and tsunami at La Palma, Canary Islands. *Geophysical Research Letters, 28,* 3397–3400.

85 Oregon Government. (2025). Cascadia subduction zone. https://www.oregon.gov/oem/hazardsprep/pages/cascadia-subduction-zone.aspx.

86 Desbruyères, D. et al. (1998). Biology and ecology of the "Pompeii worm" (*Alvinella pompejana* Desbruyères and Laubier), a normal dweller of an extreme deep-sea environment: A synthesis of current knowledge and recent developments. *Deep Sea Research Part II: Topical Studies in Oceanography, 45,* 383–422; van Dijk, P. L. M. et al. (1994). Exercise in the cold: High energy turnover in Antarctic fish. In: G. di Prisco et al., eds. *Fishes in Antarctica.* Berlin: Springer Verlag, pp. 225–236.

87 Integrated Taxonomic Information System. https://www.itis.gov.

88 See also: Rana, T. S. & Ranade, S. A. (2009). The enigma of monotypic taxa and their taxonomic implications. *Current Science, 96,* 219–229.

89 Mora, C. et al. (2011). How many species are there on Earth and in the ocean? *PLoS Biology, 9*(8), e1001127.

90 Kaiho, K. (2022). Relationship between extinction magnitude and climate change during major marine and terrestrial animal crises. *Biogeosciences, 19,* 3369–3380.

91 Longrich, N. R. et al. (2016). Severe extinction and rapid recovery of mammals across the Cretaceous–Palaeogene boundary, and the effects of rarity on patterns of extinction and recovery. *Journal of Evolutionary Biology, 29,* 1495–1512.

92 Hull, P. (2015). Life in the aftermath of mass extinctions. *Current Biology, 25,* R941–R952.

93   Hutchinson, D. K. et al. (2021). The Eocene–Oligocene transition: A review of marine and terrestrial proxy data, models and model–data comparisons. *Climate of the Past, 17*, 269–315.

94   Owen, R. (1861). *Palaeontology, or, a Systematic Study of Extinct Animals and Their Geological Relations.* Edinburgh: Adam and Charles Black; Martin, P. S. (2005). *Twilight of the Mammoths.* Berkeley, CA: University of California Press.

95   Stuart, A. J. (2005). The extinction of woolly mammoth (*Mammuthus primigenius*) and straight-tusked elephant (*Palaeoloxodon antiquus*) in Europe. *Quaternary International, 126–128*, 171–177; Webb, S. (2008). Megafauna demography and late Quaternary climatic change in Australia: A predisposition to extinction. *Boreas, 37*, 329–345.

96   Burgess, S. D. &. Bowring, S. A. (2015). High-precision geochronology confirms voluminous magmatism before, during, and after Earth's most severe extinction. *Science Advances, 1*(7), e1500470.

97   Keller, G. (2010). KT mass extinction: Theories and controversies—extended version. https://gkeller.princeton.edu/publications/kt-mass-extinction-theories-and-controversies-extended-version.

98   Raup, D. M. (1991). A kill curve for Phanerozoic marine species. *Paleobiology, 17*, 37–48.

99   Hedges, B. et al. (2015). Tree of life reveals clock-like speciation and diversification. *Molecular Biology and Evolution, 32*, 835–845.

100 Diaz, L. F. H. et al. (2019). Macroevolutionary diversification rates show time dependency. *Proceedings of the National Academy of Sciences, 116*, 7403–7408.

101 Sobel, J. M. et al. (2009). The biology of speciation. *Evolution, 64*(2), 295–315.

102 von Humboldt, A. (1808). *Ansichten der Natur mit wissenschaftlichen Erläuterungen.* Tübingen: J. G. Cotta'schen Buchhandlung.

103 Rohde, K. (1992). Latitudinal gradients in species diversity: The search for the primary cause. *Oikos, 65*, 514–527.

104 Igea, J. & Tanentzap, A. J. (2020, January 14). Angiosperm speciation speeds up near the poles. https://doi.org/10.1101/619064.

105 Melián, C. J. et al. (2012). Does sex speed up evolutionary rate and increase biodiversity? *PLoS Computational Biology, 8*(3), e1002414.

106 Momigliano, P. et al. (2017). Extraordinarily rapid speciation in a marine fish. *Proceedings of the National Academy of Sciences, 114,* 6074–6079.

107 Levy, A. A. & Feldman, M. (2022). Evolution and origin of bread wheat. *The Plant Cell, 34,* 2549–2567.

108 Wu, J. et al. (2014). Assessing and broadening genetic diversity of a rapeseed germplasm collection. *Breeding Science, 64,* 321–330.

109 Manitoba Agriculture. (2025). Canola production and management. https://www.gov.mb.ca/agriculture/crops/crop-management/canola.html.

110 Thomas, C. D. (2015). Rapid acceleration of plant speciation during the Anthropocene. *Trends in Ecology & Evolution, 30,* 448–455.

111 Otto, S. P. (2018). Adaptation, speciation and extinction in the Anthropocene. *Proceedings of the Royal Society B, 285,* 20182047.

112 Soltis, D. et al. (2004). Recent and recurrent polyploidy in *Tragopogon* (Asteraceae): cytogenetic, genomic and genetic comparisons. *Biological Journal of the Linnean Society, 82,* 485–501.

113 Thomas, C. D. (2015). Rapid acceleration of plant speciation during the Anthropocene. *Trends in Ecology & Evolution,* 30, 448–455.

114 International Union for Conservation of Nature and Natural Resources. (2025). The IUCN Red List of Threatened Species. https://www.iucnredlist.org.

115 McCallum, M. L. (2015). Vertebrate biodiversity losses point to a sixth mass extinction. *Biodiversity Conservation, 24,* 2497–2519.

116 Glazko, V. & Nei, M. (2003). Estimation of divergence times for major lineages of primate species. *Molecular Biology and Evolution, 20,* 424–434.

117 Won, Y. & Hey, J. (2005). Divergence population genetics of chimpanzees. *Molecular Biology and Evolution, 22,* 297–307.

118 Püschel, H. P. et al. (2021). Divergence-time estimates for hominins provide insight into encephalization and body mass trends in human evolution. *Nature Ecology & Evolution, 5*(6), 808–819.

119 Hartmann, P. et al. (1994). Normal weight of the brain in adults in relation to age, sex, body height and weight. *Pathologe, 15,* 165–170.

120 Hublin, J.-J. et al. (2017). New fossils from Jebel Irhoud, Morocco and the pan-African origin of *Homo sapiens. Nature, 546,* 289–292.

121 Benton, M. J. & Donoghue, P. C. J. (2007). Paleontological evidence to date the tree of life. *Molecular Biology and Evolution, 24,* 26–53.

122 Smith, G. E. (1924). *Evolution of Man: Essays.* Oxford: Oxford University Press.

123 Aiello, L. C. & Wheeler, P. (1995). The expensive-tissue hypothesis. *Current Anthropology, 36,* 199–221.

124 Studies of chimpanzee hunting and meat eating include: Wrangham, R. W. & Riss, E. Z. B. (1990). Rates of predation on mammals by Gombe chimpanzees, 1972–1975. *Primates, 31,* 157–170; Watts, D. P. et al. (2012). Diet of Chimpanzees (*Pan troglodytes schweinfurthii*) at Ngogo, Kibale National Park, Uganda, 1. Diet composition and diversity. *American Journal of Primatology, 74,* 114–129; Piel, A. K. et al. (2017). The diet of open-habitat chimpanzees (*Pan troglodytes schweinfurthii*) in the Issa valley, western Tanzania. *Journal of Human Evolution, 112,* 57–69; Moore, J. et al. (2017). Chimpanzee vertebrate consumption: Savanna and forest chimpanzees compared. *Journal of Human Evolution, 112,* 30–40.

125 Wrangham, R. (2009). *Catching Fire: How Cooking Made Us Human.* New York: Basic Books; Cornélio, A. M. et al. (2016). Human brain expansion during evolution is independent of fire control and cooking. *Frontiers in Neuroscience, 10,* 167.

126 Dunbar, R. I. M. & Shultz, S. (2017). Why are there so many explanations for primate brain evolution? *Philosophical Transactions of the Royal Society B, 372,* 20160244; Shultz, S. et al. (2012). Hominin cognitive evolution: Identifying patterns and processes in the fossil and archaeological record. *Philosophical Transactions of the Royal Society B, 367,* 2130–2140.

127 Du, A. et al. (2018). Pattern and process in hominin brain size evolution are scale-dependent. *Proceedings of the Royal Society B, 285,* 20172738.

2. Organisms: Life's strategies and capabilities

1 Mora, C. et al. (2011). How many species are there on Earth and in the ocean? *PLoS Biology, 9*(8), e1001127.

2 Gaillard, J.-M. et al. (2005). Generation time: A reliable metric to measure life-history variation among mammalian populations. *American Naturalist, 116,* 119–123.

3 Biewener, A. A. (2003). *Animal Locomotion*. Oxford: Oxford University Press.

4 Scott, G. R. et al. (2015). How bar-headed geese fly over the Himalayas. *Physiology*, *30*, 107–115; Hypoxico. (2023). Altitude to oxygen chart. https://hypoxico.com/pages/altitude-to-oxygen-chart.

5 Gerringer, M. E. et al. (2017). *Pseudoliparis swirei* sp. nov.: A newly-discovered hadal snailfish (Scorpaeniformes: Liparidae) from the Mariana Trench. *Zootaxa*, *4358*, 161–177; Blue Robotics. (2023). Pressure-to-depth and depth-to-pressure calculator. https://bluerobotics.com/learn/pressure-depth-calculator.

6 England, J. L. (2013). Statistical physics of self-replication. *Journal of Chemical Physics*, *139*, 121923.

7 Gibson, B. et al. (2018). The distribution of bacterial doubling times in the wild. *Proceedings of the Royal Society B*, *285*, 20180789.

8 Shoemaker, W. R. et al. (2021). Microbial population dynamics and evolutionary outcomes under extreme energy limitation. *Proceedings of the National Academy of Sciences*, *118*(33), e2101691118

9 Gregory, P. (2020). *Birds of Paradise and Bowerbirds: An Identification Guide*. Princeton, NJ: Princeton University Press.

10 iNaturalist. (2025). Dasyurids. Dasyurids (Family Dasyuridae), iNaturalist. https://www.inaturalist.org/taxa/40153-Dasyuridae.

11 MacArthur, R. H. & Wilson, E. O. (1967). *The Theory of Island Biogeography*. Princeton, NJ: Princeton University Press.

12 World Wildlife Fund. (2014). African elephant. https://files.worldwildlife.org/wwfcmsprod/files/Publication/file/2dxha5r3sx_African_elephant___WWF_wildlife_and_climate_change_series.pdf.

13 Ridley, M. (1988). Mating frequency and fecundity in insects. *Biological Reviews*, *63*, 509–549.

14 Sears, M. J. et al. (2020). Prolonged and variable copulation durations in a promiscuous insect species: No evidence of reproductive benefits for females. *Behavioural Processes*, *179*, 104189.

15 Stallmann, R. & Harcourt, A. H. (2006). Size matters: The (negative) allometry of copulatory duration in mammals. *Biological Journal of the Linnean Society*, *87*, 185–193.

16 Allen, R. M. et al. (2018). Applying movement ecology to marine animals with complex life cycles. *Annual Review of Marine Science*, *10*, 19–42.

17 Lacroix, A. et al. (2023). Physiology, Menarche. Stat Pearls Publishing. https://www.ncbi.nlm.nih.gov/books/NBK470216/.

18 Gerrard, D. E. & Grant, A. L. (2007). *Principles of Animal Growth and Development*. Dubuque, IA: Kendall Hunt; Lawrence, T. L. et al., eds. (2013). *Growth of Farm Animals*. Wallingford: CABI Publishing.

19 Hutchinson, J. R. et al. (2011). A computational analysis of limb and body dimensions in *Tyrannosaurus rex* with implications for locomotion, ontogeny, and growth. *PLoS ONE, 6*(10), e26037.

20 Smil, V. (2019). *Growth: From Microorganisms to Megacities*. Cambridge, MA: MIT Press.

21 Fernandes, F. A. et al. (2022). Cross-sectional and longitudinal method for describing growth curve of rabbits. *Animal Science and Technology and Inspection of Animal Products, 74*(4); Houlihan, D. F. et al. (1988). Growth rates and protein turnover in Atlantic cod, *Gadus morhua*. *Canadian Journal of Fisheries and Aquatic Sciences, 45*(6), 951–964.

22 WHO. (2025). WHO growth standards for use in US infants and children birth to 2 years. WHO Growth Charts. https://www.cdc.gov/growthcharts/who-growth-charts.htm.

23 Bogin, B. (1999). Evolutionary perspective on human growth. *Annual Review of Anthropology, 28*, 109–153; Tanner, J. M. (1962). *Growth and Adolescence*. Oxford: Blackwell Scientific; Tanner, J. M. (2010). *A History of the Study of Human Growth*. Cambridge: Cambridge University Press.

24 WHO. (2006). *Child Growth Standards*. Geneva: WHO.

25 Zhang X. et al. (2023). Prevalence of malnutrition and its associated factors among 18,503 Chinese children aged 3–14 years. *Frontiers in Nutrition* 11;10:122879.

26 Ji, C. & Chen, T. (2008). Secular changes in stature and body mass index for Chinese youth in sixteen major cities, 1950s–2005. *American Journal of Human Biology, 20*, 530–537.

27 NCD Risk Factor Collaboration (NCD-RisC). (2016). A century of trends in adult human height. *eLife, 2016*(5), e13410.

28 Graber, E. G. (2025). Adolescent development. https://www.merckmanuals.com/en-ca/home/children-s-health-issues/growth-and-development/physical-growth-and-sexual-maturation-of-adolescents.

29 Pacifici, M. et al. (2013). Generation length for mammals. *Nature Conservation, 5*, 87–94.

30 Langergraber, K. E. et al. (2012). Generation times in wild chimpanzees and gorillas suggest earlier divergence times in great ape and human evolution. *Proceedings of the National Academy of Sciences, 109,* 15716–15721.

31 Fenner, J. N. et al. (2005). Cross-cultural estimation of the human generation interval for use in genetics-based population divergence studies. *American Journal of Physical Anthropology, 128,* 415–423.

32 Russell, N. (2002). The wild side of human domestication. *Society & Animals, 10,* 285–302.

33 Smil, V. (2013). *Should We Eat Meat?* Chichester: Wiley.

34 MacDonald, J. M. & McBride, W. D. (2009). *The Transformation of U.S. Livestock Agriculture: Scale, Efficiency, and Risks.* Washington, DC: USDA.

35 Evans, A. R. et al. (2012). The maximum rate of mammal evolution. *Proceedings of the National Academy of Sciences, 109,* 4187–4190.

36 Animalia. (2023). Mountain giant Sunda rat. https://animalia.bio/mountain-giant-sunda-rat.

37 Petralia, R. S. et al. (2014). Aging and longevity in the simplest animals and the quest for immortality. *Ageing Research Reviews, 16,* 66–82.

38 Nelson, D. R. (2002). Current status of the Tardigrada: Evolution and ecology. *Integrative and Comparative Biology, 42,* 652–659.

39 Berkel, C. & Cacan, E. (2021). Analysis of longevity in Chordata identifies species with exceptional longevity among taxa and points to the evolution of longer lifespans. *Biogerontology, 22,* 329–343.

40 Zijdeman, R. L. & Ribeiro da Silva, F. (2014). Life expectancy since 1820. In: J. L. van Zanden et al., eds. *How Was Life? Global Well-being since 1820.* Paris: OECD, pp. 101–116; World Population Review. (2023). Life expectancy by country 2023. https://worldpopulationreview.com/country-rankings/life-expectancy-by-country.

41 Levine, H. J. (1997). Rest heart rate and life expectancy. *Journal of the American College of Cardiology, 30,* 1104–1106.

42 Hansson, L.-A. & Åkesson, S., eds. (2014). *Animal Movement Across Scales.* Oxford: Oxford University Press.

43 Schmidt-Nielsen, K. (1972). Locomotion: Energy cost of swimming, flying, and running. *Science, 177,* 222–228.

44 Bejan, A. et al. (2018). The fastest animals and vehicles are neither the biggest nor the fastest over lifetime. *Scientific Reports, 8,* 12925.

45 Hirt, M. R. et al. (2017). A general scaling law reveals why the largest animals are not the fastest. *Nature Ecology & Evolution, 1*, 1116–1122.

46 Videler, J. J. (1993). *Fish Swimming*. New York: Chapman and Hall.

47 Federation of American Societies for Experimental Biology. (2014, April 27). Mite sets new record as world's fastest land animal. https://www.sciencedaily.com/releases/2014/04/140427191124.htm.

48 Alexander, R. M. (1990). Size, speed and buoyancy adaptations in aquatic animals. *American Zoologist, 30*, 189–196.

49 Olla, B. L. et al. (1975). Swimming speeds of Atlantic mackerel, *Scomber scombrus*, under laboratory conditions: relation to capture by trawling. International Commission for the Northwest Atlantic Fisheries Serial No. 4039. https://www.nafo.int/Portals/0/PDFs/icnaf/docs/1976/res-76-143.pdf.

50 Shadwick, R. E. (2005). How tunas and lamnid sharks swim: An evolutionary convergence. *American Scientist, 93*, 524–531.

51 Hirt, et al. (2017). A general scaling law reveals why the largest animals are not the fastest. *Nature Ecology & Evolution, 1*, 1,116–1,122.

52 Svendsen, M. B. S. et al. (2016). Maximum swimming speeds of sailfish and three other large marine predatory fish species based on muscle contraction time and stride length: A myth revisited. *Biology Open, 5*, 1415–1419.

53 Block, B. A. et al. (1992). Direct measurement of swimming speeds and depth of blue marlin. *The Journal of Experimental Biology, 166*, 267–284.

54 Williams, R. & Noren, D. P. (2008). Swimming speed, respiration rate, and estimated cost of transport in adult killer whales. *Marine Mammal Science, 25*, 327–350. https://www.oceansinitiative.org/wp-content/uploads/2010/11/williamsnoren2009_costoftransport.pdf.

55 Tennekes, H. (2009). *The Simple Science of Flight*. Cambridge, MA: MIT Press.

56 Horvitz, N. & Sapir, N. (2014). The gliding speed of migrating birds: Slow and safe or fast and risky? *Ecology Letters, 17*, 670–679.

57 Bruderer, B. & Boldt, A. (2001). Flight characteristics of birds: I. radar measurements of speeds. *Ibis, 143*, 178–204; Alerstam, T. et al. (2007). Flight speeds among bird species: Allometric and phylogenetic effects. *PLoS Biology, 5*(8), e197.

58 Baird, E. et al. (2005). Visual control of flight speed in honeybees. *The Journal of Experimental Biology*, *208*, 3895–3905.

59 Hirt, et al. (2017).

60 Hedenström, A. et al. (2016). Annual 10-month aerial life phase in the common swift *Apus apus. Current Biology*, *26*, 3066–3070.

61 Cheney, J. A. et al. (2021). Raptor wing morphing with flight speed. *Journal of the Royal Society Interface*, *18*, 20210349.

62 Tucker, V. A. (1998). Gliding flight: Speed and acceleration of ideal falcons during diving and pull out. *The Journal of Experimental Biology*, *201*, 403–414.

63 Hart, L. A. et al. (2018). Hunting flight speeds of five southern African raptors. *Ostrich*, *89*, 251–258.

64 Hutchinson, J. R. et al. (2003). Are fast-moving elephants really running? *Nature*, *422*, 493–494.

65 Alexander, R. M. (2003). *Principles of Animal Locomotion*. Princeton, NJ: Princeton University Press; Alexander, R. M. (2005). Models and the scaling of energy cost of locomotion. *The Journal of Experimental Biology*, *208*, 1645–1652.

66 Garland, T. J. (1983). The relation between maximal running speed and body mass in terrestrial mammals. *Journal of Zoology*, *199*, 157–170.

67 Wynn, M. L. et al. (2015). Running faster causes disaster: Trade-offs between speed, manoeuvrability and motor control when running around corners in northern quolls (*Dasyurus hallucatus*). *The Journal of Experimental Biology*, *218*, 433–439.

68 Wheatley, R. et al. (2015). How fast should an animal run when escaping? An optimality model based on the trade-off between speed and accuracy. *Integrative and Comparative Biology*, *55*, 1166–1175.

69 Wilson, R. P. et al. (2015). Mass enhances speed but diminishes turn capacity in terrestrial pursuit predators. *eLife*, *2015*(4), e06487.

70 Siyabona Africa. (2024). Antelopes. https://www.krugerpark.co.za/Kruger_National_Park_Wildlife-travel/explore-kruger-park-buck-and-antelope.html.

71 Sharp, N. C. C. (1997). Timed running speed of a cheetah (*Acinonyx jubatus*). *Journal of Zoology*, *241*, 493–494; Quirke, T. et al. (2013). A comparative study of the speeds attained by captive cheetahs during the enrichment practice of the "Cheetah Run." *Zoo Biology*, *32*, 490–496.

72 Hudson, P. E. et al. (2011). Functional anatomy of the cheetah (*Acinonyx jubatus*) hindlimb. *Journal of Anatomy*, *218*, 363–374; Hudson, P. E. et al. (2012). High speed galloping in the cheetah (*Acinonyx jubatus*) and the racing greyhound (*Canis familiaris*): Spatio-temporal and kinetic characteristics. *The Journal of Experimental Biology*, *215*, 2425–2434.

73 Taylor, C. R. & Rowntree, V. J. (1973). Temperature regulation and heat balance in running cheetahs: A strategy for sprinters? *American Journal of Physiology*, *224*, 848–851.

74 Eaton, R. L. (1974). *The Cheetah*. New York: Van Nostrand Reinhold.

75 Wilson, J. W. et al. (2013). Cheetahs, *Acinonyx jubatus*, balance turn capacity with pace when chasing prey. *Biology Letters*, *9*, 20130620.

76 Wilson et al. (75); see also: Wilson, A. M. et al. (2013). Locomotion dynamics of hunting in wild cheetahs. *Nature*, *498*, 185–189.

77 Clemente, C. J. & Wilson, R. S. (2016). Speed and maneuverability jointly determine escape success: Exploring the functional bases of escape performance using simulated games. *Behavioral Ecology*, *27*, 45–54.

78 Zeder, M. A. (2011). The origins of agriculture in the Near East. *Current Anthropology*, *52* (Supplement 4), S221–S235.

79 Librado, P. et al. (2021). The origins and spread of domestic horses from the Western Eurasian steppes. *Nature*, *598*, 634–640.

80 Marks, O. (2008, December 6). The WWII German Army was 80% horse drawn; business lessons from history. https://www.zdnet.com/article/the-wwii-german-army-was-80-horse-drawn-business-lessons-from-history/.

81 Hyland, A. (1990). *Equus: The Horse in the Roman World*. New Haven, CT: Yale University Press; Busby, D. & Rutland, C. (2019). *The Horse: A Natural History*. Princeton, NJ: Princeton University Press.

82 del Castillo, B. D. (2012). *The True History of the Conquest of New Spain*. Trans: J. Burk and T. Humphrey. Indianapolis, IN: Hackett Publishing.

83 Horse Racing Sense. (2025). The Largest Horse Breeds: Shire, Belgian, Clydesdale, & More. https://horseracingsense.com/what-are-the-worlds-largest-horse-breeds/.

84 Cronin, D. et al., eds. (2001). *Irish Fairs and Markets: Studies in Local History*. Dublin: Four Courts Press; Utsav. (2023). Chetak Festival. https://utsav.gov.in/view-event/chetak-festival-1.

85 Food in Japan. (2021). Basashi. https://foodinjapan.org/kyushu/kumamoto/basashi/.

86 Koselleck, R. (2003). Der Aufbruch in die Moderne oder das Ende des Pferdezeitalters. In: *Historikerpreis der Stadt Munster 2003*. Munster: Presse- und Informationsamt.

87 Raulff, U. (2017). *Farewell to the Horse: The Final Century of Our Relationship*. London: Allen Lane.

88 Robilliard, J. J. et al. (2007). Gait characterisation and classification in horses. *The Journal of Experimental Biology, 210*, 187–197.

89 Morrice-West, A. V. et al. (2021). Variation in GPS and accelerometer recorded velocity and stride parameters of galloping Thoroughbred horses. *Equine Veterinary Journal, 53*, 1063–1074.

90 Mitsuda, T. (2007). The horse in European history, 1550–1900. PhD thesis, University of Cambridge. https://doi.org/10.17863/CAM.16030.

91 Nielsen, B. D. et al. (2006). Racing speeds of Quarter Horses, Thoroughbreds and Arabians. *Equine Veterinary Journal, 38*(S36), 128–132.

92 Vergara-Hernandez, F. B. et al. (2022). Average stride length and stride rate of Thoroughbreds and Quarter Horses during racing. *Translational Animal Science, 6*, 1–6.

93 Mercier, Q. & Aftalion, A. (2020). Optimal speed in Thoroughbred horse racing. *PLoS ONE, 15*(12), e0235024.

94 Denny, M. W. (2008). Limits to running speed in dogs, horses and humans. *The Journal of Experimental Biology, 211*, 3836–3849.

95 Guinness World Records. (2025). Fastest speed for a racehorse. https://www.guinnessworldrecords.com/world-records/fastest-speed-for-a-race-horse.

96 Sri Chinmoy. (2025). The Sri Chinmoy Self-Transcendence 3100 Mile Race. https://www.srichinmoy.org/service/sri_chinmoy_marathon_team/3100_mile_race.

97 Jordan, K. & Newell, K. M. (2008). The structure of variability in human walking and running is speed-dependent. *Exercise and Sport Science Reviews, 36*, 200–204.

98 Weyand, P. G. et al. (2010). The biological limits to running speed are imposed from the ground up. *Journal of Applied Physiology, 108*, 950–961.

99 Mehmet, H. et al. (2020). Assessment of gait speed in older adults. *Journal of Geriatric Physical Therapy, 43*, 42–52; Bohannon, R. W. & Andrews, W.

(2011). Normal walking speed: A descriptive meta-analysis. *Physiotherapy*, *97*, 182–189; Peel, N. M. (2013). Gait speed as a measure in geriatric assessment in clinical settings: A systematic review. *The Journals of Gerontology: Series A*, *68*, 39–46.

100 Olympic Games Sydney. (2000). Results. https://www.olympics. com/en/olympic-games/sydney-2000/results.

101 Majumdar, A. S. & Robergs, R. A. (2011). The science of speed: Determinants of performance in the 100 m sprint. *International Journal of Sports Science & Coaching*, *6*(3), 479–493.

102 Wilson, R. P. et al. (2015). Mass enhances speed but diminishes turn capacity in terrestrial pursuit predators. *eLife*, *4*, e06487.

103 Haugen, T. et al. (2015). 9.58 and 10.49: Nearing the *citius* end for 100 m? *International Journal of Sports Physiology and Performance*, *10*, 269–272.

104 Nag, U. (2020). Usain Bolt's records: best strikes from the Lightning Bolt. https://www.olympics.com/en/news/usain-bolt-record-world-champion-athlete-fastest-man-olympics-sprinter-100m-200m.

105 Taylor, M. J. D. & Beneke, R. (2012). Spring mass characteristics of the fastest men on Earth. *International Journal of Sports Medicine*, *33*, 667–670.

106 Čoh, M. et al. (2018). Kinematics of Usain Bolt's maximal sprint velocity. *Kinesiology*, *50*, 172–180.

107 Sterken, E. (2001). *Endurance and age: Evidence from long-distance running data*. Groningen: University of Groningen.

108 Carrier, D. R. (1984). The energetic paradox of human running and hominid evolution. *Current Anthropology*, *25*, 483–495; Bramble, D. M. & Lieberman, D. E. (2004). Endurance running and the evolution of *Homo*. *Nature*, *432*, 345–352.

109 Minetti, A. E. et al. (2002). Energy cost of walking and running at extreme uphill and downhill slopes. *Journal of Applied Physiology*, *93*, 1039–1046; Steudel-Numbers, K. L. et al. (2009). Optimal running speed and the evolution of hominin hunting strategies. *Journal of Human Evolution*, *56*, 355–360.

110 Torii, M. (1995). Maximal sweating rate in humans. *Journal of Human Ergology*, *24*, 137–152; Taylor, N. A. S. & Machado-Moreira, C. A. (2013). Regional variations in transepidermal water loss, eccrine sweat gland density, sweat secretion rates and electrolyte composition in resting and exercising humans. *Extreme Physiology & Medicine*, *2*, 1–29.

111 Lieberman, D. E. (2015). Human locomotion and heat loss: An evolutionary perspective. *Comprehensive Physiology, 5*, 99–117.

112 Almond, C. S. D. et al. (2005). Hyponatremia among runners in the Boston Marathon. *New England Journal of Medicine, 352*, 1550–1556.

113 Heinrich, B. (2001). *Racing the Antelope: What Animals Can Teach Us About Running and Life.* New York: HarperCollins; Liebenberg, L. (2006). Persistence hunting by modern hunter-gatherers. *Current Anthropology, 47*, 1017–1026.

114 Liebenberg, L. (2008). The relevance of persistence hunting to human evolution. *Journal of Human Evolution, 55*, 1156–1159.

115 Seethapathi, N. & Srinivasan, M. (2019). Step-to-step variations in human running reveal how humans run without falling. *eLife, 8*, e38371.

116 Radford, P. F. & Ward-Smith, A. J. (2003). British running performances in the eighteenth century. *Journal of Sports Sciences, 21*, 429–438.

117 Lippi, G. et al. (2008). Updates on improvement of human athletic performance: Focus on world records in athletics. *British Medical Bulletin, 87*, 7–15; Desgorces, F-D. et al. (2012). Similar slow down in running speed progression in species under human pressure. *Journal of Evolutionary Biology, 25*, 1792–1799; Denny, M. W. (2008). Limits to running speed in dogs, horses and humans. *The Journal of Experimental Biology, 211*, 3836–3849.

118 Ryder, H. J. et al. (1976). Future performance in footracing. *Scientific American, 234*(6), 109–119; Whipp, B. J. & Ward, S. A. (1992). Will women soon outrun men? *Nature, 355*, 25; Joyner, M. J. et al. (2011). The two-hour marathon: Who and when? *Journal of Applied Physiology, 110*, 275–277.

119 Marc, A. et al. (2014). Marathon progress: Demography, morphology and environment. *Journal of Sports Sciences, 32*, 524–532.

120 Wilber, R. L. & Pitsiladis, Y. P. (2012). Kenyan and Ethiopian distance runners: What makes them so good? *International Journal of Sports Physiology and Performance, 7*, 92–102.

121 Rüst, C. A. et al. (2013). Analysis of performance and age of the fastest 100-mile ultra-marathoners worldwide. *Clinics, 68*, 605–611.

122 Weyand, P. G. et al. (2010). The mass-specific energy cost of human walking is set by stature. *The Journal of Experimental Biology, 213*, 3972–3979.

123 Kelly, L. A. et al. (2018). The energetic behaviour of the human foot across a range of running speeds. *Scientific Reports, 8,* 10576.

124 Toussaint, H. M. et al. (2003). *Wave drag in front crawl swimming.* Amsterdam: Vrije Universiteit.

125 Vogt, P. et al. (2013). Analysis of 10 km swimming performance of elite male and female open-water swimmers. *SpringerPlus, 2,* 603.

126 Formenti, F. & Minetti, A. E. (2007). Human locomotion on ice: The evolution of ice-skating energetics through history. *The Journal of Experimental Biology, 210,* 1825–1833.

127 Stier, R. (2022). Kjeld Nuis clocks 103kph to smash his own speed skating record. https://www.redbull.com/se-en/kjeld-nuis-speed-skating-record-103-kph.

128 Formenti, F. et al. (2005). Human locomotion on snow: Determinants of economy and speed of skiing across the ages. *Proceedings of the Royal Society B, 272,* 1561–1569.

129 Huebsch, T. (2018, February 15). How much faster is speed skating than running? *Canadian Running Magazine.* https://runningmagazine.ca/the-scene/olympic-speedskating-running-comparison/.

130 https://www.vasaloppet.se/en/.

131 Etusuora. (2025). Holmenkollen cross-country skiing World Cup 2025. https://etusuora.com/en/cross-country-skiing/world-cup/holmenkollen.

132 AS. (2016, March 28). Ivan Origone sets new world speed skiing record: 255 km/h. https://en.as.com/en/2016/03/28/other_sports/1459177579_828788.html.

133 Whitt, F. R. & Wilson, D. G. (1982). *Bicycling Science: Ergonomics and Mechanics.* Cambridge, MA: MIT Press.

134 Cycl Endurance. (2025). The Evolution of the Average Speed of Professional Cyclists. https://4endurance.com/blogs/research/the-evolution-of-the-average-speed-of-professional-cyclists#:~:text=Today%2C%20the%20stage%20races%20regularly,few%20kilometers%20per%20hour%20more.

135 Hollingum, B. (2018). A history of cycling speed records as Denise Mueller-Korenek reaches 183 mph. Guinness World Records. https://www.guinnessworldrecords.com/news/2018/9/a-history-of-cycling-speed-records-as-denise-mueller-koronek-reaches-183-mph-541481.

136 Jackson, S. & Schouten, P. (2012). *Gliding Mammals of the World*. Melbourne: CSIRO Publishing.

137 Moncourtois, A. (2020). Gossamer Condor—the first of its kind. https://www.avinc.com/images/uploads/news/Gossamer_Condor_Use_Case.pdf.

138 AeroVironment. (2023). The flight of the Gossamer Albatross. https://www.avinc.com/about/gossamer-albatros; McIntyre, J. (1988). Man's greatest flight. https://web.mit.edu/drela/Public/web/hpa/SG_HPAG_daedalus.pdf.

3. *Speed Limits in Premodern Societies*

1 The voluminous literature on foraging societies and activities has been reviewed and evaluated in: Cummings, V. et al., eds. (2014). *The Oxford Handbook of the Archaeology and Anthropology of Hunter-Gatherers*. Oxford: Oxford University Press; Kelly, R. L. (2013). *The Lifeways of Hunter-Gatherers*. Cambridge: Cambridge University Press.

2 Obviously, large ships (from ancient Greek biremes and triremes, with three banks of oars, to large medieval and early modern galleys) manned by as many as 170 men could go faster, but with typical speeds no higher than 18 km/h (5 m/s) they were still no faster than a trotting horse. Everything else being equal, the only effective option to increase paddling speed is to reduce the drag by lightening the overall load.

3 The tanning of animal skins was perhaps the most elaborate task. Its indigenous North American version starts with flesh removal and soaking in water, followed by scraping, wringing out moisture, applying braining solution (deer brains) to treat the material, and wringing the skin again, followed by repeating the last two steps, then softening by stretching and a final smoking. For those eager to try, or just curious, there are many instructional videos on the Web.

4 Sahlins, M. (1968). Notes on the original affluent society. In: R. B. Lee and I. DeVore, eds. *Man the Hunter*. New York: Aldine Publishing, pp. 85–89.

5 Kaplan, D. (2000). The darker side of the original affluent society. *Journal of Anthropological Research*, 56, 301–324.

6 Low densities prevailed in the biosphere's extreme environments—the coldest (Arctic) and the warmest. In the latter case, population densities were low not only in arid subtropical environments but also in the densest tropical rainforests.

7 Ermenc, J. J. (1956). The Machine of Marly. *The French Review*, 29, 242–244.

8 Buringh, E. & van Zanden, J. L. (2009). Charting the "Rise of the West": Manuscripts and printed books in Europe, a long-term perspective from the sixth through eighteenth centuries. *The Journal of Economic History*, 69, 409–445.

9 Mazoyer, M. & Roudart, L. (2006). *A History of World Agriculture*. London: Earthscan.

10 Very similar modern planting and harvesting calendars for scores of crops and for all nations are available at: FAO. (2023). Crop calendar. https://cropcalendar.apps.fao.org/#/home.

11 Traditional Chinese cropping, especially in the south, was complicated by many cultivated species being grown in rotation as well as by intercropping cereals and legumes.

12 Smil, V. (2013). *Should We Eat Meat?* Chichester: Wiley, p. 58.

13 In 2023, the largest conflict zone stretched across Africa from Mali to Eritrea, with South Sudan and Somalia particularly affected by endless violence.

14 Heather, P. (2006). *The Fall of the Roman Empire*. Oxford: Oxford University Press.

15 Squatriti, P. (2019). Rye's rise and Rome's fall: Agriculture and climate in Europe during Late Antiquity. In: A. Izdebski and M. Mulryan, eds. *Environment and Society in the Long Late Antiquity*. Amsterdam: Brill, pp. 342–351.

16 Will, P.-É. et al. (1991). *Nourish the People: The State Civilian Granary System in China, 1650–1850*. Ann Arbor, MI: University of Michigan Press.

17 Pryor, F. L. (1985). The invention of the plow. *Comparative Studies in Society and History*, 27, 727–743.

18 Nag, P. K. (2011). Manual operations in farming. https://www.iloencyclopaedia.org/part-x-96841/agriculture-and-natural-resources-based-industries/farming-systems/item/538-manual-operations-in-farming.

19 Smil (2013), pp. 66–76.

20 US Department of Agriculture. (1932). *Plowing with Moldboard Plows.* Washington, DC: USDA.

21 Columella, L. J. M. (1941). *On Agriculture.* Trans. H. B. Ash. Cambridge, MA: Harvard University Press; White, K. D. (1970). *Roman Farming.* Ithaca, NY: Cornell University Press; White, K. D. (1975). *Farm Equipment of the Roman World.* Cambridge: Cambridge University Press.

22 Geza, M. (1999). Harnessing techniques and work performance of draft horses in Ethiopia. In: P. Starkey and P. Kaumbutho, eds. *Meeting the Challenges of Animal Traction.* London: Intermediate Technology Publications, pp. 143–147.

23 Raepsaet, G. (2008). Land transport, part 2: Riding, harnesses, and vehicles. In: J. P. Oleson, ed. *The Oxford Handbook of Engineering and Technology in the Classical World.* Oxford: Oxford University Press, pp. 580–605; Gans, P. J. (2004). The medieval horse harness: Revolution or evolution? A case study in technological change. In: M.-T. Zenner, ed. *Villard's Legacy: Studies in Medieval Technology, Science and Art in Memory of Jean Gimpel.* London: Routledge, pp. 175–187.

24 Bailey, L. H., ed. (1908). *Cyclopedia of American Agriculture.* New York: Macmillan.

25 Amtec. (2022). A brief history of the seed drill. https://amtec-group.com/blog/a-brief-history-of-the-seed-drill; Aldrich, L. J. (2002). *Cyrus McCormick and the Mechanical Reaper.* Greensboro, NC: Morgan Reynolds.

26 Smil (2013), pp. 105 and 109; Smil, V. (2022). *How the World Really Works.* London: Viking, pp. 48–51.

27 But by 1900 the Great Plains farming was a hybrid system, powered by animate energies but greatly benefitting from many new mechanical designs (made with newly affordable steel) as well as from the first applications of steam engines.

28 Buck, J. L. (1937). *Land Utilization in China.* Shanghai: The Commercial Press.

29 Smil (2017), p. 111.

30 Herring, M. (2020, May 24). Harvest equipment: A brief history of the combine. https://ironsolutions.com/a-brief-history-of-the-combine/.

31 Smil (2017), pp. 146–155.

32 Smil (2017), pp. 157–163.

33 Wulff, H. E. (1966). A postscript to Reti's notes on Juanelo Turriano's water mills. *Technology and Culture*, 7, 398–401.

34 Smeaton, J. (1759). An experimental enquiry concerning the natural powers of water and wind to turn mills, and other machines, depending on a circular motion. *Philosophical Transactions of the Royal Society*, 51, 100–174; Denny, M. (2004). The efficiency of overshot and undershot waterwheels. *European Journal of Physics*, 25, 193–202.

35 Oleson, J. P., ed. (2008). *The Oxford Handbook of Engineering and Technology in the Classical World*. Oxford: Oxford University Press.

36 Evans, O. (1795). *The Young Mill-Wright and Miller's Guide*. Philadelphia, PA: Oliver Evans.

37 George Washington's Mount Vernon. (2025). Ten facts about the gristmill. https://www.mountvernon.org/the-estate-gardens/gristmill/ten-facts-about-the-gristmill/.

38 Apuleius, L. *Metamorphoses IX*, 12.

39 Betz, A. (1926). *Wind-Energie und ihre Ausnutzung durch Windmühlen*. Göttingen: Vandenhoeck & Ruprecht.

40 Stokhuyzen, F. (1961). *The Dutch Windmill*. Bussum: C. A. J. Dishoeck.

41 Hollandlandofwater.com. (2025). Schermer polder. https://www.hollandlandofwater.com/schermer-polder/.

42 Baker, T. L. 2985. *A Field Guide to American Windmills*. Norman, OK: University of Oklahoma Press.

43 Tourism Shikoku. (2025). Henro, Shikoku Pilgrimage. https://shikoku-tourism.com/en/shikoku-henro/shikoku-henro#:~:text=The%20Shikoku%2088%20Temple%20Pilgramage,time%20during%20the%209th%20Century.

44 Vive el Camino. (2025). Camino de Santiago. https://vivecamino.com/en/camino-de-santiago-from/roncesvalles/.

45 Drake, B. L. (2017). Changes in North Atlantic oscillation drove population migrations and the collapse of the Western Roman Empire. *Scientific Reports*, 7, 1227.

46 Krey, A. (1921). *The First Crusade*. Princeton: Princeton University Press.

47 van Tilburg, C. (2007). *Traffic and Congestion in the Roman Empire*. London: Routledge.

48 Jackson, P. (2018). *The Mongols and the West, 1221–1410*. London: Routledge, pp. 257–261.

49 Marco Polo. (1938). *The Description of the World*. Trans. A. C. Moule and P. Pelliot. London: George Routledge and Sons.

50 Lemcke, L. (2016). *Imperial Transportation and Communication from the Third to the Late Fourth Century: The Golden Age of the Cursus Publicus*. Brussels: Éditions Latomus.

51 Scheidel, W. & Meeks, E. (2025). Orbis: The Stanford Geospatial Network Model of the Roman World. https://orbis.stanford.edu.

52 Procopius. *History of the Wars* V, 14, 6.

53 Herodotus. *Histories* VIII, 98. https://sarata.com/history/herodotus/book.8.37.html.

54 Minetti, A. E. (2003). Efficiency of equine express postal systems. *Nature, 426*, 785–786.

55 Scheidel, W. & E. Meeks. (2025).

56 Atwood, C. P. (trans.) (2023). *The Secret History of the Mongols*. London: Penguin, pp. 156–157; Silverstein, A. (2007). *Postal Systems in the Pre-Modern Islamic World*. Cambridge: Cambridge University Press.

57 Hafen, L. R. (1926). *The Overland Mail, 1849–1869: Promoter of Settlement, Precursor of Railroads*. Cleveland, OH: A. H. Clark.

58 Cottrell, W. D. (2011). On the history of the Pony Express and extending its operation into the telegraph era. *Transportation Research Record: Journal of the Transportation Research Board, 2238*, 44–49.

59 Cody, W. F. (1879). *An Autobiography*. Hartford, CT: Frank E. Bliss.

60 Legends of America. (2022). Pony Bob Haslam & the Longest Ride. https://www.legendsofamerica.com/we-ponybobhaslam/.

61 Muybridge, E. (1878). The horse in motion. "Sallie Gardner," owned by Leland Stanford; running at a 1:40 gait over the Palo Alto track, 19th June 1878 / Muybridge. https://www.loc.gov/item/97502309/.

62 Selgin, G. & Turner, J. L. (2011). Strong steam, weak patents, or the myth of Watt's innovation-blocking monopoly, exploded. *The Journal of Law & Economics, 54*, 841–861.

63 Casson, L. (1951). Speed under sail of ancient ships. *Transactions of the American Philological Association, 82*, 136–148; Whitewright, J. (2011). The potential performance of ancient Mediterranean sailing rigs. *International Journal of Nautical Archaeology, 40*, 2–17.

64 de Chateaubriand, F.-R. (1850). *Memoirs from beyond the Grave 1800–1815.* Trans. A. Andriesse. New York: New York Review of Books (2022), pp. 219–220.

65 Anderson, R. & Anderson, R. C. (1927). *The Sailing Ship: Six Thousand Years of History.* London: George Harrap; Chapelle, H. I. (1967). *The Search for Speed under Sail 1700–1855.* New York, W. W. Norton; Anderson, B. D. (2008). The physics of sailing. *Physics Today*, 61, 38–43.

66 Frame, W. & Walker, L. (2018). *James Cook: The Voyages.* Montreal: McGill-Queens University Press.

67 Carter, W. E. &. Carter, M. S. (2010). The Age of Sail: A time when the fortunes of nations and lives of seamen literally turned with the winds their ships encountered at sea. *The Journal of Navigation*, 63, 717–731; Baron, J. H. (2009). Sailors' scurvy before and after James Lind—a reassessment. *Nutrition Reviews*, 67, 315–332.

68 Chatterton, E. K. (1926). *The Ship Under Sail.* London: Fisher Unwin.

69 Coleridge, S. T. (1798). *The Rime of the Ancient Mariner* (text of 1834). https://www.poetryfoundation.org/poems/43997/the-rime-of-the-ancient-mariner-text-of-1834.

70 Maury, M. F. (1855). *The Physical Geography of the Sea.* New York: Harper & Brothers, p. 171. See also: J. W. Wayland (1930). *The Pathfinder of the Seas: The Life of Matthew Fontaine Maury.* Richmond: Garrett & Massie.

71 Kelly, M. & Gráda, C. Ó. (2014). *Speed under sail, 1750–1850.* Dublin: UCD Centre for Economic Research Working Paper Series, No. WP 14/10.

72 Solar, P. M. & Hens, L. (2016). Ship speeds during the Industrial Revolution: East India Company ships, 1770–1828. *European Review of Economic History*, 20, 66–78.

73 Bruijn, J. R. et al. (1987). *Dutch-Asiatic Shipping in the 17th and 18th Centuries.* The Hague: Martinus Nijhoff.

74 Clark, A. H. (1911). *The Clipper Ships Era.* New York: G. P. Putnam's Sons; McKay, R. C. (1928). *Donald McKay and His Famous Sailing Ships.* New York: G. P. Putnam's Sons; Ross, D. G. (2012). *The Era of the Clipper Ships.* Lexington, KY: D. G. Ross III.

75 Shaw, D. (2000). *Flying Cloud: The True Story of America's Most Famous Clipper Ship and the Woman Who Guided Her.* New York: William Morrow.

76 MacGregor, D. R. (1993). *British and American Clippers: A Comparison of their Design, Construction and Performance*. London: Conway Maritime Press Limited.

77 AP. (2013, February 16). Yacht sets record from New York to San Francisco. https://www.usatoday.com/story/news/nation/2013/02/16/yacht-sets-record-from-new-york-to-san-francisco/1925311/.

78 Marine Insight. (2012, December). The guide to slow steaming on ships. https://www.marineinsight.com/wp-content/uploads/2013/01/The-guide-to-slow-steaming-on-ships.pdf.

79 Zahid, H. J. et al. (2016). Agriculture, population growth, and statistical analysis of the radiocarbon record. *Proceedings of the National Academy of Sciences*, 113, 931–935.

80 Gignoux, C. R. et al. (2011). Rapid, global demographic expansions after the origins of agriculture. *Proceedings of the National Academy of Sciences*, 108, 6044–6049.

81 Cartier, M. (2002). La population de la Chine au fil des siècles. In: I. Attané, ed., *La Chine au seuil de XXIe siècle: Questions de population, questions de société*. Paris: Institut National d'Études Démographiques, pp. 21–31.

82 US Census Bureau. (2022). Historical Estimates of World Population. https://www.census.gov/data/tables/time-series/demo/international-programs/historical-est-worldpop.html.

83 Maddison, A. (2007). *Contours of the World Economy, 1–2030 AD*. Oxford: Oxford University Press.

84 Broadberry, S. et al. (2018). China, Europe and the Great Divergence: A study in historical national accounting, 980–1850. Journal of *Economic History*, 78, 955–1000. https://www.cambridge.org/core/journals/journal-of-economic-history/article/abs/china-europe-and-the-great-divergence-a-study-in-historical-national-accounting-9801850/6451E62524E28874293D8ED6DED9A24F.

85 Smil (2013), pp. 218–219, 368–369.

86 On cathedral construction, see: Fitchen, J. (1997). *The Construction of Gothic Cathedrals: A Study of Medieval Vault Erection*. Chicago: University of Chicago Press; Jenkins. S. (2022). *Cathedrals: Masterpieces of Architecture, Feats of Engineering, Icons of Faith*. New York: Rizzoli.

87 Our World in Data. (2025). Share of the population living in urbanized areas. https://ourworldindata.org/grapher/long-term-urban-population-region.

88 de Vries, J. & van der Woude, A. (1997). *The First Modern Economy: Success, Failure, and Perseverance of the Dutch Economy, 1500–1815*. Cambridge: Cambridge University Press; Broadberry, S. et al. (2015).

89 Smil (2017), p. 368.

90 Segedunum. (2025). Catapults. https://segedunumromanfort.org.uk/catapults.

91 Miller, D. P. et al. (2019). The ballistics of seventeenth century musket balls. *Journal of Conflict Archaeology*, 14, 25–36.

92 de Ségur, P. (1825). *History of the Expedition to Russia Undertaken by the Emperor Napoleon in the Year 1812*. London: Treuttel & Würtz; Foord, E. (1914). *Napoleon's Russian Campaign of 1812*. London: Hutchinson & Company.

93 Napoleon.org. (2025). Napoleon's Russian campaign: From the Niemen to Moscow. https://www.napoleon.org/en/history-of-the-two-empires/timelines/napoleons-russian-campaign-from-the-niemen-to-moscow/; RouteYou. (2025). Napoleon's march to Moscow. https://www.routeyou.com/en-ru/route/view/391569.

4. *The Fast Lane: How modern societies accelerate*

1 Remarkably, large galleys (common warships of the antiquity and Middle Ages) were used in battle until the end of the 18th century, most notably during the Second Battle of Svensksund in July 1790, when Swedish naval forces defeated the Russian fleet. Åkesson, P. (1998). The Swedish–Russian sea battles of 1790. https://www.abc.se/~pa/mar/russ1790.htm.

2 Kuhn, O. (2004). Ancient Chinese drilling. *CSEG Recorder*, 29, 39–43.

3 Designs of these machines evolved, but the principle of *magna rota* (the treadmill crane) remained the same: men treading inside large wooden wheels. Compare the carving from the Roman tomb of Haterii from the late first century CE (https://www.museivaticani.va/content/museivaticani/en/collezioni/musei/museo-gregoriano-profano/

Mausoleo-degli-Haterii.html) with Pieter Bruegel's 1563 painting of the Tower of Babel (https://www.bruegel2018.at/en/the-tower-of-babel/).

4 Anderson, K. J. (1991). A history of lubricants. *MRS Bulletin*, *16*, 69.

5 Roux, V. & Jeffra, C. (2015). The spreading of the potter's wheel in the ancient Mediterranean. A social context-dependent phenomenon. In: W. Gauss et al., eds. *The Transmission of Technical Knowledge in the Production of Ancient Mediterranean Pottery*. Vienna: Österreichisches Archäologisches Institut, pp. 165–182.

6 Stokhuyzen, F. (1962). *The Dutch Windmill*. Bussum: C. A. J. Dishoeck.

7 Jones, R. V. (1970). The "plain story" of James Watt. *Proceedings of the Royal Society A*, *316*, 449–471.

8 Watt, J. (1781). *Specification of Patent, October 25th, 1781, for Certain New Methods of Applying the Vibrating or Reciprocal Motion of Steam or Fire Engines, to Produce a Continued Rotative or Circular Motion Round an Axis or Centre, and Thereby to Give Motion to the Wheels of Mills or Other Machines*. In: J. P. Muirhead. (1854). *The Origins and Progress of the Mechanical Inventions of James Watt* (Vol. 3). London: John Murray.

9 SteamLocomotive.com. (2025). Steam locomotive drive wheel types. https://www.steamlocomotive.com/types/drivers/.

10 Reynolds, J. (1970). *Windmills and Watermills*. London: Hugh Evelyn.

11 Barlow, J. (1791). Conflagration or the merry mealmongers. The British Museum, Prints and Drawings. https://www.britishmuseum.org/collection/object/P_1880-1113-1491.

12 Blake, W. (1810). Jerusalem. https://www.poetryfoundation.org/poems/54684/jerusalem-and-did-those-feet-in-ancient-time.

13 Musson, A. E. (1978). *The Growth of British Industry*. New York: Holmes & Meier, p. 41.

14 Daugherty, C. R. (1933). Horsepower equipment in the United States, 1869–1929. *The American Economic Review*, *23*, 428–440.

15 Historic UK. (2025). Rainhill Trials – Historic UK. https://www.historic-uk.com/HistoryUK/HistoryofBritain/Rainhill-Trials/.

16 Rhodes, J. T. &. Stephenson, D. R. (2013). Steam locomotive rail wheel dynamics part i: Precedent speed of steam locomotives. https://static1.squarespace.com/static/55e5ef3fe4b0d3b9ddaa5954/t/55e63647e4b06159647b2fb9/1441150535974/WP_SLRWD_1.pdf.

17 Preble, G. H. (1883). *A Chronological History of the Origin and Development of Steam Navigation*. Philadelphia: L. R. Hamersly & Co.

18 Burgh, N. P. (1869). *A Practical Treatise on Modern Screw Propulsion*. London: E. and F. N. Spon; Shipping Wonders of the World. (2023). Development of the screw propeller. https://www.shippingwondersoftheworld.com/screw_propeller.html.

19 Nature. (1937, February 13). Centenary of practical marine screw propulsion. *Nature, 139*, 279.

20 Halpern, S. (2015). Speed and revolutions. https://www.titanicology.com/Titanica/SpeedandRevolutions.htm.

21 The Shipyard. (2020). Propeller cavitation explained. https://www.theshipyardblog.com/propeller-cavitation-explained/.

22 Croft, T. (1922). *Steam Engine Principles and Practice*. New York: McGraw-Hill.

23 Smil, V. (2005). *Creating the Twentieth Century*. New York: Oxford University Press.

24 Parsons, C. A. (1911). *The Steam Turbine*. Cambridge: Cambridge University Press.

25 Parsons, R. H. (1936). *The Development of Parsons Steam Turbine*. London: Constable & Company.

26 Siemens Energy. (2025). Reliable steam turbines for improved efficiency. https://www.siemens-energy.com/global/en/home/products-services/product-offerings/steam-turbines.html.

27 Langen, A. (1919). *Nicolaus August Otto, der Schöpfer des Verbrennungsmotors*. Stuttgart: Franck; Sittauer, H. L. (1972). *Gebändigte Explosionen*. Berlin: Transpress Verlag für Verkehrswesen.

28 Daugherty. (1933).

29 Diesel, R. (1913). *Die Entstehung des Dieselmotors*. Berlin: Julius Springer.

30 Smil, V. (2010). *Prime Movers of Globalization*. Cambridge, MA: MIT Press.

31 Wärtsilä. (2025). RTA and RT-flex low-speed engines. https://www.wartsila.com/marine/services/2-stroke-engine-services/rta-and-rt-flex-low-speed-engines.

32 Tesla, N. (1888). *Electro-magnetic Motor. Specification forming part of Letters Patent No. 391,968, dated May 1, 1888*. Washington, DC: US Patent Office. https://www.uspto.gov.

33 Neidhöfer, G. (2008). *Michael von Dolivo-Dobrowolsky und der Drehstrom.* Berlin: VDE Verlag.

34 The Engineering Toolbox. (2025). Electrical motors. https://www.engineeringtoolbox.com/speed-electrical-motors-d_738.html.

35 Drives & Controls. (2009). 104,000 rpm appliance motor is "world's fastest". https://drivesncontrols.com/104000-rpm-appliance-motor-is-worlds-fastest/; Live Science. (2008, November 14). New spin record set: 1 million rpm. https://www.livescience.com/3075-spin-record-set-1-million-rpm.html.

36 New Steam Age. (1942). First practical gas turbines. *New Steam Age* 1(1), 9–10, 20; ASME. (1988). The world's first industrial gas turbine set at Neuchâtel (1939). New York: ASME; Smil, V. (2006). *Transforming the Twentieth Century.* New York: Oxford University Press.

37 Rolls-Royce. (2015). *The Jet Engine.* Chichester: Wiley.

38 Pile Buck International. (2009). *Pile Driving by Pile Buck.* Vero Beach, FL: Pile Buck International.

39 Kumar, E. A. (2023). Lathe machine. https://themechanicalengineering.com/lathe-machine/.

40 Moxon. J. (1701). *Mechanick Exercises: Or, the Doctrine of Handy-Works Applied to the Art of Smithing, Joinery, Carpentry, and Turning.* London: J. Moxon.

41 Diderot, D. & D'Alembert, J. L. R. (1751–1777). *L'Encyclopédie ou Dictionnaire Raisonné des Sciences, des Arts et des Métiers.* Paris: Avec Approbation et Privilège du Roy.

42 Gilbert, K. R. (1971). *Henry Maudslay: Machine Builder.* London: HM Stationery Office.

43 Mumford, L. (1934). *Technics and Civilization.* London: Routledge & Kegan Paul, p. 80.

44 For example, modern jetliners require the first of four periodic checks after 600–1,000 hours of flying, the second after 6–8 months, the third after less than two years, and the most thorough one after 6–10 years in service: Kinnison, H. & Siddiqui, T. (2011). *Aviation Maintenance Management.* New York: McGraw-Hill.

45 De-icing usually takes less than 10 minutes and its effect lasts for up to 22 minutes: if dangerous conditions persist and an aircraft does not depart before that time, it must be de-iced again.

46 You can see total delays (within, into, or out of the United States) and cancellations in real time at: https://www.flightaware.com/live/cancelled/.

47 Hughes, M. (1988). *Rail 300: The World High Speed Train Race.* Newton Abbot: David and Charles.

48 Smith, R. A. (2004). Railway speed-up: A review of its history, technical developments and future prospects. *JSME International Journal, 47,* 444–450.

49 American Society of Mechanical Engineers. (1980). *The Pioneer Zephyr.* https://www.asme.org/wwwasmeorg/media/resourcefiles/aboutasme/who%20we%20are/engineering%20history/landmarks/58-pioneer-zephyr-1934.pdf.

50 Smil, V. (2014). Fifty years of *Shinkansen. The Asia-Pacific Journal, 12,* 1–9.

51 Japan Rail Pass. (2025). Shinkansen: The Japanese bullet trains. https://www.jrailpass.com/shinkansen-bullet-trains.

52 Nippon.com. (2018). The Tōkaidō Shinkansen's world-class safety, reliability, and frequency. https://www.nippon.com/en/features/h00201/.

53 Arduin, J.-P. & Ni, J. (2005, March). French TGV network development. *Japan Railway & Transport Review,* 22–28.

54 Lawrence, M. et al. (2019). *China's High-Speed Rail Development.* Washington, DC: World Bank. https://documents1.worldbank.org/curated/en/933411559841476316/pdf/Chinas-High-Speed-Rail-Development.pdf.

55 Global Times. (2019). China develops new traction motor for 400 km/h high-speed trains. https://www.globaltimes.cn/content/1164757.shtml.

56 Amtrak. (2025). Acela train. https://amtrakguide.com/routes/acela-express/.

57 ShipIndex.org. (2025). SS Royal William. Shipindex.org. https://www.shipindex.org/vessels/Q612004.

58 Doe, H. (2017). *The First Atlantic Liner: Brunel's Great Western Steamship.* Stroud: Amberley Publishing.

59 Great Ocean Liners. (n.d.). The Blue Riband. https://www.greatoceanliners.com/blue-riband.

60 QueenMary.com. (n.d.). Final years at sea. https://www.queenmary.com/timeline-stats--fun-facts.htm.

61 Royal Caribbean. (n.d.). Icon of the Seas. https://www.royalcarib-bean.com/cruise-ships/icon-of-the-seas.

62 Cudahy, B. J. (2006). *Box Boats: How Container Ships Changed the World.* New York: Fordham University Press; Smil (2010).

63 MAN Energy Solutions. (2018). *Basic Principles of Ship Propulsion.* Copenhagen: MAN Energy Solutions; Ng, M. W. (2019). Vessel speed optimisation in container shipping: A new look. *Journal of the Operational Research Society*, 70, 541–547.

64 Yang, H. & Xing, Y. (2020). Containerships sailing speed and fleet deployment optimization under a time-based differentiated freight rate strategy. *Journal of Advanced Transportation*, 2020, 4103275.

65 Lindstad, H. & Eskeland, G. S. (2015). Low carbon maritime transport: How speed, size and slenderness amounts to substantial capital energy substitution. *Transportation Research Part D: Transport and Environment*, 41, 244–256.

66 Finnsgård, C. et al. (2020). The shipper's perspective on slow steaming—Study of Six Swedish companies. *Transport Policy*, 86, 44–49.

67 Freightcourse. (2023). Port cranes. https://www.freightcourse.com/port-cranes/.

68 Miller, G. (2022, November 23). Zero ships waiting off Southern California for first time since 2020. *Freight Waves*. https://www.freightwaves.com/news/zero-ships-waiting-off-southern-california-59-off-other-ports.

69 United Nations Conference on Trade and Development. (2020). *Review of Maritime Transport 2019*. New York: UNCTAD.

70 World Bank. (2025). Container port traffic. https://data.worldbank.org/indicator/IS.SHP.GOOD.TU.

71 Walz, W. & Niemann, H. (1997). *Daimler-Benz: Wo das Auto Anfing.* Konstanz: Verlag Stadler; Flower, R. & Jones, M. W. (1981). *100 Years of Motoring*. Maidenhead: McGraw-Hill Book Company.

72 Mercedes-Benz. (2023). First Mercedes Model Series 1900–1901. https://mercedes-benz-publicarchive.com/marsClassic/en/instance/ko/First-Mercedes-model-series-19001901.xhtml?oid=5899.

73 Hanlon, M. (2017). The fastest cars in history: 1894 to 1914. https://newatlas.com/worlds-fastest-production-cars-1894-1914/46196/.

74 Federal Highway Administration. (2017). The *Reichsautobahnen*. https://highways.dot.gov/highway-history/interstate-system/reichsautobahnen.

75 Federal Highway Administration. (2012). History of the Interstate Highway System. https://highways.dot.gov/highway-history/interstate-system/50th-anniversary/history-interstate-highway-system.

76 AutoEurope. (2025). Driving on the Autobahn in Germany: Must-read tips and info. https://www.autoeurope.ca/travel-guides/germany/driving-the-autobahn-in-germany/.

77 Evans, S. et al. (2024). The 25 fastest cars in the world, ranked. https://robbreport.com/motors/cars/gallery/fastest-production-cars-1234810377/.

78 Kennedy, D. (2020). 470 MPH! Speed Demon is the fastest piston-driven car on Earth. https://www.nhra.com/news/2020/470-mph-speed-demon-fastest-piston-driven-car-earth.

79 Mihalascu, D. (2023). Pininfarina Battista is world's fastest-accelerating road-legal car. https://insideevs.com/news/622936/pininfarina-battista-is-worlds-fastest-accelerating-road-legal-car/.

80 Reference. (2015, August 4). Is NASCAR the number one spectator sport? https://www.reference.com/world-view/nascar-number-one-spectator-sport-6b86eeefa976ec33.

81 Formula 1. (2023). F1 Schedule 2023. https://www.formula1.com/en/racing/2023.

82 For a selection of this popular genre see: https://www.imdb.com/list/ls033275519/.

83 Global Road Safety Facility. (2012). Factsheet: The Relation Between Speed and Crashes (Institute for Road Safety Research/SWOV), GRSF. https://www.globalroadsafetyfacility.org/publications/factsheet-relation-between-speed-and-crashes-institute-road-safety-researchswov.

84 Safety Topics. (2025). Speed and Space Management. https://www.csiconcrete.com/safetytopics/speedspace.pdf.

85 University of Pennsylvania. (2025). Vehicle_Stopping_Distance.doc. https://www.decamillismattingly.com/wp-content/uploads/2020/05/vehicle_stopping_distance_and_time_upenn.pdf.

86 But modern gliders are amazing machines: they have a glide ratio (the distance an aircraft can travel forward for every unit of altitude lost during a glide) in excess of 60:1 (the Boeing 747 has a glide ratio of 15:1), which means that when they are towed to 3 kilometers above ground, they can glide for nearly 200 kilometers—and, aided by thermals, tens of hours.

87 But the exertion required to keep these light machines aloft means that they will not be available for common public usage.

88 Smil, V. (2021). *Grand Transitions: How the Modern World Was Made.* New York: Oxford University Press, pp. 25–28.

89 McCullough, D. (2015). *The Wright Brothers.* New York: Simon & Schuster.

90 Skytamer. (2018). 1913 Chronology of aviation history. https://www.skytamer.com/1913.html.

91 KLM. (2025). History of KLM. https://www.klm.com/information/corporate/history. ·

92 Swopes, B. R. (2019). Handley Page HP.42. https://www.thisdayin-aviation.com/tag/handley-page-hp-42/.

93 Pirie, G. (2009). Incidental tourism: British Imperial air travel in the 1930s. *Journal of Tourism History, 1,* 49–66.

94 Jones, G. (2014). *Douglas DC-3: 80 Glorious Years.* Stroud: Fonthill Media.

95 Dick, H. G. & Robinson, D. H. (1985). *The Golden Age of the Great Passenger Airships Graf Zeppelin & Hindenburg.* Washington, DC: Smithsonian Institution Press.

96 Grossman, D. et al. (2017). *Zeppelin Hindenburg: An Illustrated History of LZ-129.* Cheltenham: The History Press.

97 Bowers, P. M. (1982). The Boeing 314 Clipper. https://aeroresources-inc.com/uploads/198211-1939%20Boeing%20314%20Clippers.pdf.

98 Maguire, D. R. (2016). Enemy jet history. *The Aeronautical Journal, 52,* 75–84.

99 GAC speed converter. https://www.globalaircraft.org/converter.htm.

100 Eckardt, D. (2017, April 12). Breaking the magic 1000. Rolls-Royce Heritage Trust, 2017 President Evening Lecture; van der Linden, B. (2022). Breaking the sound barrier: Chuck Yeager and the Bell X-1. https://airandspace.si.edu/stories/editorial/breaking-sound-barrier-75th.

101 Military Factory. (2023). Korean War jet aircraft. https://www.military-factory.com/aircraft/korean-war-jet-aircraft.php#google_vignette.

102 National Air and Space Museum. (2023). North American F-100 Super Sabre. https://airandspace.si.edu/collection-objects/north-american-f-100d-super-sabre/nasm_A19781577000.

103 Lockheed Martin. (2025). F-22 Raptor. https://www.lockheedmartin.com/en-us/products/f-22.html; Airforce Technology. (2021). Sukhoi

Su-57—A significant boost to Russian air combat capabilities. https://www.airforce-technology.com/features/sukhoi-su-57-a-significant-boost-to-russian-air-combat-capabilities/.

104 Global Aircraft. (n.d.). F-15 Eagle. https://www.globalaircraft.org/planes/f-15_eagle.pl.

105 Simons, G. M. (2013). *Comet! The World's First Jet Airliner*. Barnsley: Penn and Sword Aviation.

106 Global Aircraft. (n.d.). Boeing 707. https://www.globalaircraft.org/planes/b707.pl.

107 Global Aircraft. (n.d.). Boeing 747. https://www.globalaircraft.org/planes/b747.pl; Modern Airliners. (2022). What are the Boeing 787 Dreamliner specs? https://www.modernairliners.com/boeing-787#specs.

108 Butcher, L. (2010). *Aviation: Concorde*. London: House of Commons Library; Glancy, J. (2016). *Concorde: The Rise and Fall of the Supersonic Airliner*. Boston: Atlantic Books; Buttler, T. & Carbonel, J.-C. (2018). *Building Concorde: From Drawing Board to Mach 2*. Forest Lake, MN: Specialty Press.

109 Tennekes, H. (2009). *The Simple Science of Flight*. Cambridge, MA: MIT Press, p. 168.

110 Wikipedia. (2023). List of F-15 losses. https://en.wikipedia.org/wiki/List_of_F-15_losses.

111 Pawlyk, O. (2017, June 17). The F-22 fighter jet restart is dead: Study. https://www.military.com/daily-news/2017/06/21/the-f22-fighter-jet-restart-dead-study.html.

112 Moreover, many of these operations on large American farms have been subcontracted to companies specializing in seeding, fertilizing, and harvesting, obviating the need to own expensive field machinery that is used for only a limited number of hours a year.

113 And even less than two seconds in some states: Smil, V. (2023). *Invention and Innovation*. Cambridge, MA: MIT Press, p. 51.

114 Lebergott, S. (1966). Labor force and employment, 1800–1960. In: D. S. Brady, ed. *Output, Employment, and Productivity in the United States after 1800*. Cambridge, MA: NBER, pp. 117–204.

115 American Society of Mechanical Engineers. (2009). *Hughes Two-Cone Drill Bit*. https://www.asme.org/wwwasmeorg/media/resourcefiles/

aboutasme/who%20we%20are/engineering%20history/landmarks/246-hughes-two-cone-drill-bit-1909.pdf.

116 Hughes, H. R. (1909). Drill. https://patents.google.com/patent/US930759A/en.

117 Garfield, L. E. & Scott, F. L. (1933). Roller bearing bit. https://patents.google.com/patent/US2030442A/.

118 Vuik, R. et al. (2010). *Percussion: Manual Drilling Series.* Papendrecht: Practica Foundation; Smil, V. (2017). *Oil.* London: Oneworld.

119 Daugherty (1933).

120 Schurr, S. H. et al. (1990). *Electricity in the American Economy: Agent of Technological Progress.* New York: Greenwood Press.

121 Cobb, H. M. (2010). *The History of Stainless Steel.* Materials Park, OH: ASM International.

122 King. R. I., ed. (1985). *Handbook of High-Speed Machining Technology.* Berlin: Springer.

123 Gtools. (2025). End Mill Feeds and Speeds: Charts & Data. https://www.6gtools.com/technical-info/end-mills/feeds-and-speeds-carbide.html?srsltid=AfmBOopC5La6T-pEx6kEp6Yza8FX4oiloxqQyYTPUr-Z3XT754ZGv5SDx.

124 Wegener, K. et al. (2016). Success story cutting. *Procedia CIRP, 46,* 512–524.

125 Dowson, D. & Hamrock, B. J. (1981). *History of Ball Bearings.* Cleveland, OH: NASA

126 Vaughan, P. (1794). Axle trees, arms, and boxes. British Patent No. 2006 of A.D. 1794, 1–2.

127 SKF. (2023). Annual Report 2022. https://investors.skf.com/sites/skf-ir/files/pr/202303013297-1.pdf.

128 Wang, L. (2021). British English-speaking speed 2020. *Academic Journal of Humanities & Social Sciences, 4,* 93–100.

129 Coupé, C. et al. (2019). Different languages, similar encoding efficiency: Comparable information rates across the human communicative niche. *Science Advances, 5*(9), eaaw2594.

130 Horne, J. et al. (2011). Computerised assessment of handwriting and typing speed. *Educational & Child Psychology, 28,* 52–66.

131 Zheng, J. & Meister, M. (2025). The unbearable slowness of being: Why do we live at 10 bits/s? *Neuron, 113,* 192–204. https://arxiv.org/pdf/2408.10234.

132 Eisenstein, E. L. (1980). *The Printing Press as an Agent of Change*. Cambridge: Cambridge University Press; Hoe, R. (1902). *A Short History of the Printing Press*. New York: Robert Hoe.

133 Standage, T. (2014.) *The Victorian Internet: The Remarkable Story of the Telegraph and the Nineteenth Century's On-line Pioneers*. New York: Bloomsbury.

134 Hertz, H. (1893). *Electric Waves: Being Researches on the Propagation of Electric Action with Finite Velocity through Space*. Trans. D. E. Jones. London: Macmillan.

135 Smith, A. & Paterson, R., eds. (1998). *Television: An International History*. Oxford: Oxford University Press.

136 Moore, G. E. (1965). Cramming more components onto integrated circuits. *Electronics, 38*(8), 114–117; Moore, G. E. (1975). Progress in digital integrated electronics. *Technical Digest, IEEE International Electron Devices Meeting*, 11–13; Moore, G. E. (2003). No exponential is forever: But "Forever" can be delayed! (Paper presented at IEEE International Solid-State Circuits Conference, San Francisco.) https://ieeexplore. ieee.org/document/1234194/.

137 Intel. 2025. Announcing a New Era of Integrated Electronics. https:// www.intel.com/content/www/us/en/history/virtual-vault/articles/ the-intel-4004.html.

138 Hafner, K. & Lyon, M. (1996). *Where Wizards Stay Up Late: The Origins of the Internet*. New York: Simon & Schuster; McCullough, B. (2018). *How the Internet Happened: From Netscape to the iPhone*. New York: Liveright.

139 Fighter Planes and Military Aircraft. (2024). Speed of sound at different altitudes and temperatures. https://fighter-planes.com/jetmach1.htm.

140 PIC Wire and Cable. (2023). Velocity factor in cables. https://www. picwire.com/Files/Technical-Articles/Velocity-Factor_PIC_Technical-Article.pdf.

141 Agrawal, G. P. (2021). *Fiber-Optic Communication Systems*. Chichester: Wiley.

142 Franklin, B. (2012). Benjamin Franklin Journal of a voyage from England to Philadelphia 1726. American History. http://www.let.rug.nl/ usa/documents/1701-1750/benjamin-franklin-journal-of-a-voyage-from-england-to-philadelphia-1726.php.

143 Bell Mobility. Speed test. https://speed.is/ca/bell/.

144 Gautier, F. (2016). *L'oeuvre de Claude Chappe: Créateur de l'Administration française des Télégraphes*. Paris: Hachette; Field, A. J. (1994). French optical telegraphy, 1793–1855: Hardware, software, administration. *Technology and Culture, 35*, 315–347.

145 History of the Atlantic Cable & Undersea Communications. (2021). Messages carried by the 1858 Atlantic telegraph cable. https://atlantic-cable.com/Article/1858Messages/index.htm.

146 Rutgers. (2023). Automatic telegraphy. https://edison.rutgers.edu/life-of-edison/inventions?view=article&id=527:automatic-telegraphy&catid=91.

147 Ifrah, G. (2002). *The Universal History of Computing*. New York: Wiley; Haigh, T. & Ceruzzi, P. E. (2021). *A New History of Modern Computing*. Cambridge, MA: MIT Press.

148 Nordhaus, W. D. (2007). Two centuries of productivity growth in computing. *The Journal of Economic History, 67*, 128–159.

149 Bell, G. (2014). *Supercomputers: The Amazing Race (A History of Supercomputing, 1960-2020)*. San Francisco: Microsoft. https://gordonbell.azurewebsites.net/MSR-TR-2015-2_Supercomputers-The_Amazing_Race_Bell.pdf; Grover, G. et al. (2018). Evolution of supercomputers. *International Journal of Computer Applications, 181*, 27–29; Hewlett-Packard. 2023. HP Cray EX Supercomputer. https://www.hpe.com/psnow/doc/a00094635enw.

150 Berndt, E. R. et al. (2000). Price and quality of desktop and mobile personal computers: A quarter century of history. http://www.nber.org/~confer/2000/si2000/berndt.pdf; https://www.singularity.com/charts/page61.html; Etiemble, D. (2018). 45-year CPU evolution: One law and two equations. http://dx.doi.org/10.48550/arXiv.1803.00254.

151 Galazzo, R. (2020). Timeline from 1G to 5G: A brief history on cell phones. https://www.cengn.ca/information-centre/innovation/timeline-from-1g-to-5g-a-brief-history-on-cell-phones/.

152 Qualcomm. (2023). Everything you need to know about 5G. https://www.qualcomm.com/5g/what-is-5g.

153 Thrift, N. (1996). *Spatial Formations*. London: Sage; Virilio, P. (2006) *Speed and Politics*. Los Angeles: Semiotext(e); Wajcman, J. & Dodd, N. (2016). *The Sociology of Speed: Digital, Organizational, and Social Temporalities*. Oxford: Oxford University Press; Harvey, D. (2008). The

condition of postmodernity. In: S. Seidman & J. C. Alexander, eds. *The New Social Theory Reader*. London: Taylor and Francis.

154 National Army Museum, (2023). Weapons of the Western front. https:// www.nam.ac.uk/explore/weapons-western-front; National WWII Museum. (2018, November 9). The MG-42 machine gun. https:// www.nationalww2museum.org/war/articles/mg-42-machine-gun.

155 Wellerstein, A. (2020). Counting the dead at Hiroshima and Nagasaki. *Bulletin of the Atomic Scientists*. https://thebulletin.org/2020/08/ counting-the-dead-at-hiroshima-and-nagasaki/.

156 Princeton University. (2019). Plan A. https://sgs.princeton.edu/the-lab/plan-a.

5. Parting Musings: Obverses, limits, aspirations

1 Grueskin, C. et al. (2020). *Safe Slaughter*. New Haven, CT: Yale University. https://law.yale.edu/sites/default/files/area/center/leap/ document/safe_slaughter_-_cafe_lab_-_spring_2020.pdf.

2 Strategic Organizing Center. (2023). *In Denial: Amazon's Continuing Failure to Fix Its Injury Crisis*. https://thesoc.org/resources/in-denial-amazons-continuing-failure-to-fix-its-injury-crisis/.

3 The Ocean Cleanup. (2023). The Great Pacific Garbage Patch. https:// theoceancleanup.com/great-pacific-garbage-patch/.

4 There is an extensive, and disheartening, literature about the failures and inadequacies of material recycling. A place to start, among many: Plastic Soup Foundation. (2023). Recycling myth. https://www.plasticsoupfoundation.org/blog/recycling-myth/; Crownhart, C. (2021, August 19). Solar panels are a pain to recycle. These companies are trying to fix that. *MIT Technology Review*. https://www.technologyreview.com/2021/08/19/1032215/solar-panels-recycling/.

5 Beaudreau, B. C. (2020). *The Economics of Speed: Machine Speed as the Key Factor in Productivity*. Cham: Springer, p. 121.

6 Bell, S. W. (2020). Horse racing in imperial Rome: Athletic competition, equine performance, and urban spectacle. *The International Journal of the History of Sport*, 37, 183–232.

7 Neumann, C. (1971). A note on Alexander's march-rates. *Historia: Zeitschrift für Alte Geschichte, 20,* 196–198.

8 Caesar, G. J. (1869). *De bello gallico.* (Gallic War). Trans. W. A. McDevitte and W. S. Bohn. New York: Harper & Brothers (book 4, chapter 17); Taylor, J. H. (1902). Caesar's Rhine bridge. *The Classical Review, 16,* 29–34.

9 Hagia Sofia's construction took place between 532 and 537, but its high central dome (48.5 meters tall and 31 meters in diameter) collapsed in 558 and was speedily rebuilt just four years later: Ousterhout, R. (1999). *Master Builders of Byzantium.* Princeton: Princeton University Press.

10 Fall, A. et al. (2004). Sliding friction on wet and dry sand. *Physics Review Letters, 112,* 175502.

11 Smil, V. (2021). *Grand Transitions: How the Modern World Was Made.* New York: Oxford University Press, pp. 25–28.

12 Beaudreau (2020), p. 5.

13 The speed of plowing is determined by tractor power, plow width, and the type of soil: each of these varies substantially and hence there are no widely representative averages.

14 Derpsch, R. et al. (2010). Current status of adoption of no-till farming in the world and some of its main benefits. *International Journal of Agricultural & Biological Engineering, 3*(1), 1–25.

15 Alan Newman (2023). Pillsbury A-Mill. *Atlas Obscura.* https://www.atlasobscura.com/places/pillsbury-a-mill; North Dakota Mill. (2023). About us. https://www.ndmill.com/about.

16 For details about subsequent waves of steelmaking, see: Smil, V. (2016). *Still the Iron Age: Iron and Steel in the Modern World.* Amsterdam: Elsevier.

17 Smil (2021).

18 Casey, R. (2008). *The Model T: A Centennial History.* Baltimore: The Johns Hopkins University Press.

19 Barlowe, A. (1898). The First Voyage to Roanoke. 1584. The First Voyage Made to the Coasts of America, with Two Barks, wherein Were Captains M. Philip Amadas and M. Arthur Barlowe, Who Discovered Part of the Countrey Now Called Virginia, anno 1584. Written by One of the Said Captaines, and Sent to Sir Walter Ralegh, Knight, at Whose Charge and Direction, the Said Voyage Was Set Forth.

20 Franklin (2012). https://www.let.rug.nl/usa/documents/1701-1750/ benjamin-franklin-journal-of-a-voyage-from-england-to-philadel-phia-1726.php.

21 Alternative Transport. (2023). Isochrone map of Berlin 1819, 1906 and 2015. https://alternativetransport.wordpress.com/2016/12/20/isochrone-map-of-berlin-1819-1906-and-2015/.

22 Paullin, C. O. & Wright, J. K. (1932). *Atlas of the Historical Geography of the United States.* https://dsl.richmond.edu/historicalatlas/.

23 Travel China Guide. (2023). Beijing–Shanghai high speed train. https://www.travelchinaguide.com/china-trains/beijing-shanghai-highspeed.htm.

24 Souza, L. (2022). United States carriers are cutting flights to regional airports. https://simpleflying.com/united-states-carriers-cutting-flights-regional-airports/.

25 Cairncross, F. (1995). The death of distance: A survey of telecommunications. *The Economist, 336*(7934), 5–28; Cairncross, F. (1997). *The Death of Distance: How the Communications Revolution Will Change Our Lives.* Boston: Harvard Business School Press.

26 Brodkin, J. (2022). 1.1 quintillion operations per second: US has world's fastest supercomputer. *Ars Technica.* https://arstechnica.com/information-technology/2022/05/1-1-quintillion-operations-per-second-us-has-worlds-fastest-supercomputer/.

27 For details, see: Smil (2019).

28 Smil, V. (2017). *Energy and Civilization.* Cambridge, MA: MIT Press, pp. 101–105.

29 International Energy Agency. (2022). World balance (2020). https://www.iea.org/data-and-statistics/data-product/world-energy-balances.

30 For sectoral shares see: World Bank. (2023). Value added (% of GDP). For agriculture the share was 4.3 percent in 2021.

31 Koinoniki-epitheorisi.gr. (2023). ΣΠΕΥΔΕ ΒΡΑΔΕΩΣ. https://www.koinoniki-epitheorisi.gr/articles/philosophy/158-speude-vradeos?start=1.

32 Bernik, R. and Vučajnk, F. (2008). Grain losses during harvesting of winter barley and winter wheat with the harvester Deutz FahrPowerliner 4035 H. https://www.cabidigitallibrary.org/doi/full/10.5555/20193168266.

33 Fitts, P. M. & Peterson, J. R. (1964). Information capacity of discrete motor responses. *Journal of Experimental Psychology, 67,* 103–112.

34 Centers for Disease Control and Prevention. (2024). Global road safety. https://www.cdc.gov/transportation-safety/global/?CDC_AAref_Val= https://www.cdc.gov/injury/features/global-road-safety/index.html.

35 National Institutes of Health. (2023). What risk factors do all drivers face? https://www.nichd.nih.gov/health/topics/driving/condition-info/risk-factors.

36 Żuchowski, A. (2016). Analysis of the influence of the impact speed on the risk of injury of the driver and front passenger of a passenger car. *Eksploatacja i Niezawodnosc—Maintenance and Reliability*, *18*, 436–444.

37 Rosén, E. et al. (2011). Literature review of pedestrian fatality risk as a function of car impact speed. *Accident Analysis & Prevention*, *43*, 25–33.

38 National Highway Traffic Safety Administration. (2025). Seat belts. https://www.nhtsa.gov/vehicle-safety/seat-belts.

39 Ding, C. et al. (2019). Motorcyclist injury risk as a function of real-life crash speed and other contributing factors. *Accident Analysis & Prevention*, *123*, 374–386.

40 Hydén, C. (2019). Speed in a high-speed society. *International Journal of Injury Control and Safety Promotion*, *27*, 44–50; Goble, K. (2021, January 27). Bills in seven states would alter speed limits. https://landline.media/bills-in-seven-states-would-alter-speed-limits/.

41 The quest for automotive speed was an important part of German history long before the mass automobilization of the post–Second World War decades, from the invention of the automobile by Benz, Daimler, and Maybach in the 1880s to Adolf Hitler's specification that Porsche must design a people's car (*Volkswagen*) capable of 100 km/h: Nelson, W. H. (1998). *Small Wonder: The Amazing Story of the Volkswagen Beetle*. Cambridge, MA: Robert Bentley.

42 Melosi, M. V. (2010). The automobile shapes the city. http://autolife.umd.umich.edu/Environment/E_Casestudy/E_casestudy1.htm.

43 The Geography of Transport Systems. (2023). Average number of hours of delay per auto commuter per year, selected American cities, 1982–2020. https://transportgeography.org/contents/chapter8/urban-transport-challenges/traffic-delays-united-states/.

44 Air Routes and Ground Services. (2023). Regional jets' airline frequencies down 33% since 2019, OAG data reveals. https://airlinergs.com/regional-jets-airline-frequencies-down-33-since-2019-oag-data-reveals/.

45 In Canada perhaps the most egregious example of relying on short-haul shuttles is the Toronto–Ottawa link: 114 weekly flights, a distance of a mere 362 kilometers (ideal for rapid trains) with a scheduled flying time of 1 hour and 3 minutes. Adding just an hour for checking in and security and just half an hour for trips to and from the airport results in an average speed of about 145 km/h!

46 National Institute of General Medical Sciences. (2023). Circadian rhythms. https://nigms.nih.gov/education/fact-sheets/Pages/circadian-rhythms/.

47 US Bureau of Labor Statistics. (2023). Injuries, illnesses, and fatalities. https://www.bls.gov/web/osh/summ1_00.htm.

48 Cook, I. (2018). How fast is too fast? OSHA's regulation of the meat industry's line speed and the price paid by humans and animals. *Sustainable Development Law & Policy, 18*(1), 39.

49 Pork Information Gateway. (2012). Repetitive motion. https://porkgateway.org/resource/repetitive-motion/; Mukhopadhyay, P. & Khan, A. (2015). The evaluation of ergonomic risk factors among meat cutters working in Jabalpur, India. *International Journal of Occupational and Environmental Health, 21*, 192–198.

50 Human Rights Watch. (2005). Worker health and safety in the meat and poultry industry. https://www.hrw.org/reports/2005/usa0105/4.htm.

51 Polansek, T. & Huffstutter, P. J. (2021). As U.S. pork plant speeds up slaughtering, workers report more injuries. Reuters. https://www.reuters.com/world/us/us-pork-plant-speeds-up-slaughtering-workers-report-more-injuries-2021-02-19/.

52 According to WHO's dashboard, more than 13.3 billion vaccines were administered by April 2023: World Health Organization. Coronavirus (COVID-19) dashboard. https://data.who.int/dashboards/covid19/cases?n=c. Accessed 2023.

53 Belliveau, N. M. et al. (2020). Fundamental limits on the rate of bacterial growth. *bioRxiv*. https://doi.org/10.1101/2020.10.18.344382.

54 Dutch men average 184 centimeters, East Timorese 159 centimeters, a difference of 25 centimeters: See: World Data. Average height for men and women worldwide. https://www.worlddata.info/average-bodyheight.php.

55 Desgorces, F-D. et al. (2008). From Oxford to Hawaii ecophysiological barriers limit human progression in ten sport monuments. *PLoS ONE, 3*(11), e3653.

56 And these spectacles have been professionalized: https://major-leagueeating.com.

57 For comparison of available growth estimates, see: US Census Bureau. (2022). Historical estimates of world population. https://www.census.gov/data/tables/time-series/demo/international-programs/historical-est-worldpop.html.

58 CIA. (2024). Total fertility rate. https://www.cia.gov/the-world-factbook/field/total-fertility-rate/country-comparison/.

59 World Population Review. Countries with declining population. https://worldpopulationreview.com/country-rankings/countries-with-declining-population. Accessed 2024.

60 Desine, S. et al. (2023). Daily step counts before and after the COVID-19 pandemic among All of Us (AOU) research participants. *JAMA Network Open, 6*(3), e233526.

61 The entire realm of domestic chores is a perfect counterexample to common claims of accelerating "progress": not one of the many electrified gadgets has seen any fundamental speed changes (just some marginal speed-ups and, uniformly, better and easier controls) since the time of its mass-scale commercial adoption that, in many cases, was two or three generations ago!

62 Hauer, T. (2015). Every society is dromocratic: Speed and philosophy. *Journal of Multidisciplinary Engineering Science and Technology, 2*(9), 2445–2448.

63 Insurance Institute for Highway Safety. Maximum posted speeds by state. https://www.iihs.org/topics/speed/speed-limit-laws. Accessed 2023.

64 Hughes, M. (2007, May 1). V150: 574·8 km/h eclipses the 1990 world record. *Railway Gazette*. https://www.railwaygazette.com/in-depth/v150-5748-km/h-eclipses-the-1990-world-record/25027.article.

65 Rivera, I. (2020). The limits of high speed rail. https://mappingignorance.org/2020/01/22/the-limits-of-high-speed-rail/.

66 Delow, P. (2011). The sound of a Shinkansen train. https://stophs2.org/news/3543-shinkansen-sound.

67 Tennekes, H. (2009). *The Simple Science of Flight*. Cambridge, MA: MIT Press.

68 For the history and prospects of SST, see: Smil, V. (2023). *Invention and Innovation*. Cambridge, MA: MIT Press.

69 US Department of Energy. (2017). How do wind turbines survive severe weather and storms? https://www.energy.gov/eere/articles/how-do-wind-turbines-survive-severe-weather-and-storms?nrg_redirect=465731.

70 The European overall wind capacity factor averages less than 30 percent, but new, large onshore projects now operate at close to 40 percent and the latest offshore farms at more than 50 percent of their capacity: Wind Europe. (2021). *Wind Energy in Europe*. https://s1.elespanol.com/2021/02/24/actualidad/210224_windeurope_combined_2020_stats.pdf.

71 Shockley, W. & Queisser, H. J. (1961). Detailed balance limit of efficiency of p-n junction solar cells. *Journal of Applied Physics*, *32*, 510–519. Germany's gloomy climate results in a solar capacity factor of just short of 11 percent, while the US average is nearly 25 percent.

72 Beaudreau (2020), p. 121.

73 Torque is measured in Newton-meters (Nm; or foot-pounds in old style). For example, torque wrenches commonly used by auto mechanics to tighten lug nuts and bolts on car suspension and engine mounts, or to change tires, have a torque between 40 and about 340 Nm: https://www.maxprocorp.com/blog/common-torque-wrench-sizes-finding-the-right-size/—while the world's largest diesel engine, powering giant container ships, has a torque of more than 7 million Nm: Puiu, T. (2023). This is what 109,000 horsepower looks like—meet the biggest and most powerful engine in the world. https://www.zmescience.com/feature-post/technology-articles/engineering/biggest-most-poweful-engine-world/.

74 There are many excellent books reviewing the progress of steam engines—a recent one is: Rosen, W. (2012). *The Most Powerful Idea in the World: A Story of Steam, Industry, and Invention*. Chicago: University of Chicago Press.

75 Dickinson, H. W. (1939). *A Short History of the Steam Engine*. Cambridge: Cambridge University Press, p. 152.

76 The station's massive presence and emissions of soot and sulfur dioxide had also greatly interfered with the nearby Greenwich Observatory. Porrino, M. (2012). Notes on technological and architectural aspects of London Transport power stations and substations, 1880–1915. *Fourth International Congress on Construction History*, Paris, 3–7 July 2012, Paris, France.

77 Diesel, R. (1913). *Die Entstehung des Dieselmotors*. Berlin: Julius Springer.

78 Wärtsilä Corporation. (2006). The world's most powerful engine enters service. https://www.wartsila.com/media/news/12-09-2006-the-world%27s-most-powerful-engine-enters-service.

79 See annual electricity generation data in the Energy Institute's *Statistical Review of World Energy*. https://www.energyinst.org/statistical-review.

80 Electric Vehicle News. (2009). Navy tests world's most powerful superconductor ship motor. http://www.electric-vehiclenews.com/2009/06/navy-tests-worlds-most-powerful.html.

81 Alstom. (1988). The world's first industrial gas turbine set—GT Neuchâtel. https://www.asme.org/wwwasmeorg/media/resourcefiles/aboutasme/who%20we%20are/engineering%20history/landmarks/135-neuchatel-gas-turbine.pdf.

82 Siemens Energy. (2023). SGT5-9000HL gas turbine. https://www.siemens-energy.com/global/en/home/products-services/product/sgt5-9000hl.html.

83 Memon, O. (2023). How much do aircraft engines weigh? *Simple Flying*. https://simpleflying.com/how-much-do-aircraft-engines-weigh/.

84 Langston, L. S. (2013). The adaptable gas turbine. *American Scientist*, *101*, 264–267.

85 National Academies of Sciences, Engineering, and Medicine. (2016). *Commercial Aircraft Propulsion and Energy Systems Research: Reducing Global Carbon Emissions*. Washington, DC: The National Academies Press. https://nap.nationalacademies.org/catalog/23490/commercial-aircraft-propulsion-and-energy-systems-research-reducing-global-carbon.

86 Data available in annual statistics published by the International Civil Aviation Organization. https://www.icao.int/sustainability/Pages/Statistics.aspx.

87 Imai, M. (1986). *Kaizen: The Key to Japan's Competitive Success*. New York: McGraw-Hill; Encyclopedia of Detroit. (2023). Ford Rouge Complex. https://detroithistorical.org/learn/encyclopedia-of-detroit/ford-rouge-complex/.

88 Beaudreau (2020), p. 75.

89 Peterson, R. (2018). Impacts of airline deregulation. https://onlinepubs.trb.org/onlinepubs/trnews/trnews315airlinedereg.pdf.

90 Gordon, R. J. (2016). *The Rise and Fall of American Growth: The U.S. Standard of Living since the Civil War*. Princeton, NJ: Princeton University Press; Erber, G. et al. (2017). The global productivity slowdown: Diagnosis, causes and remedies. *Intereconomics, 2017*, 45–50; Winkler, J. et al. (2021). Re-evaluating the sources of the recent productivity slowdown. https://cepr.org/voxeu/columns/re-evaluating-sources-recent-productivity-slowdown.

91 Keats, E. C. et al. (2019). Improved micronutrient status and health outcomes in low- and middle-income countries following large-scale fortification: Evidence from a systematic review and meta-analysis. *American Journal of Clinical Nutrition, 109*, 1696–1708.

92 In 2020, Africa had 30 state-based conflicts, 11 of them civil wars, and in 2021 countries experiencing external involvement in their domestic conflicts included Burkina Faso, Burundi, Cameroon, Central African Republic, DR Congo, Ethiopia, Kenya, Mali, Mozambique, Niger, Nigeria, and Somalia: Palik, J. et al. (2022). *Conflict Trends in Africa, 1989–2021*. Oslo: Peace Research Institute of Oslo.

93 For maglev history and the latest news, see: https://phys.org/tags/maglev/.

94 Tesla (2013). Hyperloop Alpha. https://www.tesla.com/sites/default/files/blog_images/hyperloop-alpha.pdf; Smil (2023).

95 Smil (2023), pp. 102–105.

96 Boom Supersonic. (2022). Boom—Supersonic Passenger Airplanes. https://boomsupersonic.com/.

97 O'Kane, S. (2017). Elon Musk proposes city-to-city travel by rocket, right here on Earth. https://www.theverge.com/2017/9/29/16383048/elon-musk-spacex-rocket-transport-earth-travel.

98 Powell, C. S. (2017). Elon Musk's wild idea for city-to-city rockets just might work. https://www.nbcnews.com/mach/science/elon-musk-s-wild-idea-city-city-rockets-just-might-ncna812386.

Coda

1 Bouskill, K. E. et al. (2018). *Speed and Security*. Santa Monica, CA: Rand Corporation, pp. 1–2.

2 Bouskill et al. (2018), p. 1.

3 Colvile, R. (2016). *The Great Acceleration: How the World is Getting Faster, Faster*. New York: Bloomsbury.

4 Markov, I. L. (2014). Limits on fundamental limits to computation. *Nature*, 512, 147–154.

5 While many British and American publications of the latter half of the 19th century wrote admiringly about the latest technical achievements, already in 1848 Karl Marx and Friedrich Engels, in their *Communist Manifesto*, pointed out "the burden of toil also increases, whether by prolongation of the working hours, by increase of the work exacted in a given time or by increased speed of the machinery"—and Marx, in *Das Kapital* (published in 1867), made many repeated references to the increased speed of machinery as a tool of exploitation. And Wajcman and Dodd entitled their introductory chapter to their book about the sociology of speed "The powerful are fast, the powerless are slow": Wajcman, J. & Dodd, N. (2016). Introduction: The powerful are fast, the powerless are slow. In: *The Sociology of Speed: Digital, Organizational, and Social Temporalities*. Oxford: Oxford University Press, p. 1.

6 American Stroke Association. (2023). Acute ischemic stroke. https://www.stroke.org/-/media/Stroke-Files/Ischemic-Stroke-Professional-Materials/AIS-Toolkit/AIS-Professional-Education-Presentation-ucm_485538.

7 Most of the more than a billion people who must rely on seasonal interior heating can now just set the thermostats governing their natural gas furnaces, electric heaters, or heat pumps. Compare that speed with the speed of securing wood (gathering or buying it), storing it, chopping it, feeding the wood stoves, and removing ashes (and still facing the risk of CO poisoning). Over a lifetime, these time savings add up to months of avoided chores.

8 Warf, B. (2008). *Time-Space Compression: Historical Geographies*. London: Routledge; Wajcman, J. (2008). Life in the fast lane? Towards a sociology of technology and time. *The British Journal of Sociology*, 59(1), 59–77.

9 Andrews, G. et al. (2022). Speed and space: Rates of motion in health and wellbeing. *Wellbeing, Space and Society*, *3*, 100112.

10 Moreover, commercial flying now combines near-sonic speed with increasingly common flight delays and cancellations, with passengers forced to spend hours in overcrowded airports (some overnight on benches or floor)—and there is no prospect for any early improvement: Stock, S. et al. (2023, July 25). Flight delays, cancellations could continue for a decade. CBS News. https://www.cbsnews.com/news/the-future-of-flying-more-delays-more-cancellations-more-chaos/.

11 Slow Food International. (2023.) Our history. https://www.slowfood.com/our-history/; Tranter, P. & Tolley, R. (2020). *Slow Cities: Conquering our Speed Addiction for Health and Sustainability*. Amsterdam: Elsevier.

12 Macrotrends. (2023). McDonalds's Gross Profit 2010–2023. https://www.macrotrends.net/stocks/charts/MCD/mcdonalds/gross-profit. Accessed 2024.

13 Illich, I. (1974). *Energy and Equity*. New York: Harper & Row, p. 16.

14 Illich (1974), pp. 18–19. As we have seen, the average walking speed (depending on age and terrain) is 3–4 mph rather than the 4.7 mph implied by Illich.

15 US data for 2022–2023 (accessed 2024): average cost of a new car $48,000, https://www.cbsnews.com/video/average-new-car-cost-us-48000-one-model-selling-less-than-20000/; average life of a vehicle 11 years, https://www.caranddriver.com/research/a32758625/how-many-miles-does-a-car-last/; average distance driven 21,687 km/year, https://www.fhwa.dot.gov/ohim/onh00/bar8.htm; average after-tax earnings $28/hour, https://www.ziprecruiter.com/Salaries/After-Tax-Salary; average cost of gasoline $2,000/year, https://www.mach1services.com/how-much-spend-on-fuel-each-year/; average cost of insurance $2,150/year, https://www.forbes.com/advisor/car-insurance/average-cost-of-car-insurance/; average cost of maintenance $800/year, https://www.aaa.com/autorepair/articles/what-does-it-cost-to-own-and-operate-a-car.

16 National Highway Traffic Safety Administration. (2023, January 10). Traffic crashes cost America $340 billion in 2019. https://www.nhtsa.gov/press-releases/traffic-crashes-cost-america-billions-2019.

17 Prottoy, A. A. et al. (2023). *The Fast, the Slow, and the Congested: Urban transportation in rich and poor countries.* Cambridge, MA: National Bureau of Economic Research.

18 Li, A. (2021). *Decoding China's Car Industry: 40 Years.* Singapore: World Scientific.

19 Boulding, K. E. (1974). The social system and the energy crisis. *Science, 184,* 255–257.

20 Starr, C. (1969). Social benefit versus technological risk. *Science, 165,* 1232–1238.

21 Fieselmann, J. et al. (2022). What are the reasons for refusing a COVID-19 vaccine? A qualitative analysis of social media in Germany. *BMC Public Health, 22,* 846; Sekizawa, Y. et al. (2022). Association between COVID-19 vaccine hesitancy and generalized trust, depression, generalized anxiety, and fear of COVID-19. *BMC Public Health, 22,* 126.

22 Fischoff, B. et al. (1978). How safe is safe enough? A psychometric study of attitudes towards technological risks and benefits. *Policy Sciences, 9,* 127–152; Slovic, P. (1987). Perception of risk. *Science, 236,* 280–285; Slovic, P. (2000). *The Perception of Risk.* London: Earthscan.

23 Illich (1974), p. 18; Airbus. (2023). Global market forecast. https:// www.airbus.com/en/products-services/commercial-aircraft/ global-market-forecast.

24 Illich (1974).

25 On cobalt: Murray, A. (2022, September 27). Cobalt mining: The dark side of the renewable energy transition. Earth.org. https://earth.org/ cobalt-mining. On vegetables grown in plastic tunnels: Galloway, H. (2021, August 12). Europe's vegetable garden is ridden with poverty, plastic and contradiction. Euronews. https://www.euronews.com/ my-europe/2021/08/12/europe-s-vegetable-garden-is-ridden-with- poverty-plastic-and-perversity.

26 Christie, N. et al. (2023). Delivering hot food on motorcycles: A mixed method study of the impact of business model on rider behaviour and safety. *Safety Science, 158,* 105991.

27 For faster aircraft boarding methods see: Steffen, J. H. & Hotchkiss, J. (2012). Experimental test of airplane boarding methods. *Journal of Air Transport Management, 18,* 64–67; Delcea, C. et al. (2019). Methods

for accelerating the airplane boarding process in the presence of apron buses. *IEEE Access*, 17, 134372–134387.

28 Lord Byron. (1819). *Don Juan*. https://www.gutenberg.org/files/21700/21700-h/21700-h.htm.

29 Henley, W. E. (1919). *Poems*. New York: Charles Scribner's Sons, pp. 207, 220–221.

30 And before you might dismiss the author as another privileged Victorian, learn about his life of suffering and determination. Diniejko, A. (2011). William Ernest Henley: A biographical sketch. https://victorianweb.org/authors/henley/introduction.html.

31 Barthes, R. (1957). The New Citroën. In: *Mythologies*. Trans: A. Lavers. New York: Noonday Press, pp. 88–89.

32 Smil, V. (2025). *Creating and Transforming the Twentieth Century*. New York: Oxford University Press.

Index